Thomas Edison, Chemist

Thomas
EDISON, CHEMIST

American
Chemical
Society

Washington, D.C.

1971

BYRON M. VANDERBILT

Library of Congress Catalog Card 75-172526
ISBN 8412-0129-3
PRINTED IN THE UNITED STATES OF AMERICA

Contents

Preface

I HAVE LONG been interested in Thomas Edison's outstanding record of technical accomplishments. However, it was not until late 1968, when I had an opportunity to study his work in more detail, that I realized he had made major contributions in the field of applied chemistry and chemical engineering. Here was a story largely untold. This book is an attempt to tell that story.

Many of the books on Edison have been of the hero-worship type. I have tried to report Edison's errors and failures as well as his successes. However, because of the satisfaction of describing a man whose approach to research and development was so daring—even when spending his own money—and his faith in the experimental method so complete, it has been difficult to remain completely unprejudiced.

This book is directed towards the general reader. In the case of each subject discussed, background information is given on the state of the art at the time Edison did his work. I have attempted to present the chemical technology involved in such a way so that those having no formal training or experience in chemistry can understand what Edison was doing. For the reader who is a professional chemist or chemical engineer, I hope this book will broaden his background in the history of chemistry.

Chemistry is the most basic of all sciences; it deals with the compositions and properties of materials. In Edison's time it was taught primarily as a supporting science in such fields as medicine, agriculture, and mineralogy. Edison was one of the first to grasp the importance of chemical knowledge when conducting industrial research and development. There was more chemical know-how employed in the development of the incandescent electric light than there was knowledge of electricity and electrical engineering. Once the basic principle of the phonograph was discovered, the years of arduous work on its development by Edison and others dealt largely with the composition and handling of materials for records—again a chemical problem.

When Edison began his adult life, the United States was a sparsely settled nation. The mass of its population was ill-housed and ill-fed; railroads and the telegraph had provided little in the way of improved transportation and communication for the average citizen; most important, the long hours of harsh labor in the homes, on farms, and in factories left little time for the niceties of life. Edison, probably more than any other single individual, created devices which made man's life more pleasant. Although environmental problems, congested cities, and crowded highways have resulted from our labor-saving devices and overall affluence, to an American working 12 hours a day, seven days a week, such matters would have had little meaning.

In this book there is no attempt to explain how Edison was able to develop so many new things and introduce them into the marketplace. Such an analysis is complex. I hope to publish something on Edison, the innovator, at a future date.

In writing a book of this type one must obviously rely heavily on published literature. Although an attempt has been made to read all available books dealing with Edison's life, his many patents and the various articles on his work in scientific journals have been my principal sources of technical information. Archival information has also been obtained from the various historic sites of the Edison era, as well as from companies closely associated with Edison or now manufacturing products he initiated. These companies are too numerous to mention individually. Some will be apparent from citations in the section dealing with references and appendices at the end of the book. To say that many people have been graciously helpful in locating materials for this book is certainly an understatement.

Although one hesitates to mention any specific names since so many others should be mentioned, special thanks are due Dr. Loren G. Polhamus of Chevy Chase, Maryland, formerly of the U.S. Department of Agriculture, who looked over the rough draft of Chapter 9 and offered several helpful comments. Mr. Edward F. Schweitzer, Alkaline Battery Division of ESB Incorporated, kindly read and commented on the draft of Chapter 7. Mr. Harold S. Anderson, Assistant Curator of the Edison National Historic Site, West Orange, New Jersey, and Mr. Norman R. Speiden, former curator at the West Orange Site now living in Lakewood, New Jersey, both read over the final manuscript. Finally, various staff members of General Electric Company's Research and Development Center checked the manuscript for technical accuracy. The American Chemical Society joins me in thanking these authorities for checking in detail the historical and technical aspects of this book.

Very special recognition is given Mrs. Mildred L. Winters of Westfield, New Jersey, who not only handled the stenographic work relating to this treatise but also assisted in the search of material and offered many helpful editorial suggestions.

BYRON M. VANDERBILT

Scotch Plains, New Jersey
April 1971

, |THOMAS ALVA EDISON was born February 11, 1847 and died October 18, 1931. Of his life span of nearly 85 years, 73 of them were working years. Considering that his idea of an eight-hour working day was eight hours in the forenoon and eight hours in the afternoon and that vacations, if any, were strictly of the working type, Edison's productive life span was probably twice that of a typical present-day chemist.

Thomas was born in Milan, Ohio, a thriving town on the Huron River, eight miles from Lake Erie. In 1829 the Milan founding fathers had built a canal extending northward three miles to the navigable part of the river. Until the coming of the railroad in 1853, Milan was the strategic inland port for shipment of farm produce. During the wheat harvesting season as many as 300 wagons—many of huge size drawn by six-horse teams—arrived daily by dirt roads from a maximum distance of about 150 miles. Lake steamers and schooners were towed through the canal to a basin which was 600 feet long and 250 feet wide. Fourteen warehouses lined the banks of the basin, as well as a shipbuilding yard. Grain left Milan for foreign ports *via* Lake Erie, the Erie Canal, and the Hudson River. Milan was also an overnight stop for wagon trains headed farther west, including those going to California induced by the discovery of gold there in 1848. Thus, young Tom, or Al as he was usually called in his childhood, spent his early years in an exciting environment of industry and commerce.

1

From 1830 to 1855 Milan had a very interesting history, and a visit to Edison's birthplace and the nearby Milan Historical Museum is well worth one's time. The town is located two miles south of the Ohio Turnpike. The handsome brick museum is the former home of Dr. Lehman Galpin, who assisted at Thomas' birth. The Edison home was built in the early 1840's on the side of a hill overlooking the basin where the ships loaded and turned around for the return trip. Once the lake shore railroad was established, most farmers found markets for their grain and lumber closer than Milan. Business at this inland port dropped precipitously. Most of the warehouses at the canal basin were torn down. In 1867 a flood washed out the two sets of canal locks and the docks in the basin. The canal was abandoned, and now tall trees and brush grow where commerce once thrived below the birthplace of Thomas Edison.

When the new railroad by-passed Milan on its way to Toledo, Sam Edison, Tom's father, quickly realized that his shingle business was doomed. In 1854 the family moved to Port Huron, Michigan. Port Huron is across the St. Clair River from Sarnia, Ontario, just below where the river leaves Lake Huron on its way to Lake St. Clair. In 1814 Captain Gratiot of the U.S. Army built a fort on the west bank, 20 feet above water, about 1,000 feet below Lake Huron. Its small cannon could control passage in this strategic waterway, and the fort served as protection for the settlers in the area from the Indians. Fort Gratiot was maintained by the Army until 1879.

The Edisons purchased a two-story colonial style house on the southern edge of the Military Reservation, formerly occupied by the post storekeeper. The new home had spacious grounds of 10 acres, and Tom soon had an opportunity to learn the chores connected with vegetable and fruit growing and a cow to provide milk for the family. Valuable white pine and oak trees lined the lower Lake Huron and stretched along the St.

Thomas Edison's birthplace, Milan, Ohio, as it appears today

Clair River. The Black River, which empties into the St. Clair at Port Huron, brought vast supplies of logs to be processed by the saw mills. Thus, with the bustling lumber industry, river traffic, a military camp nearby, and a foreign country within eyesight, there was much to stir the imagination of an alert boy. Tom had been a rather sickly child and shortly after the move to Port Huron became seriously ill with scarlet fever. This delayed his formal education, and it was not until the fall of 1855, when at the age of eight and one-half, he was enrolled in a one-room school. Tom had already shown that he was a difficult child in that he was continually plaguing his parents and neighbors with questions of the why- and how-type. He proved to be completely incompatible with the methods used in conventional

3

education at that time; after three months his formal education came to an end. It appears to have been by mutual agreement. Tom ran off from school one day and refused to return. The schoolmaster felt the child was too stupid to learn, his mother decided she wanted to teach him herself, and the father was either having financial difficulty or was reluctant to pay the small school fee. A schoolteacher before marriage, his mother was a gentle person. Prior to Tom's birth she had lost two of her younger children, and Eliza, age three, died 10 months after Tom's birth. These tragedies brought her closer to Tom, her seventh and last child; she no doubt had a highly protective attitude towards him and understood the boy much better than his father. Realizing that Tom had unusual reasoning powers, his mother ignored the ABC approach and began at once to read world history and the classics to him. The boy became fascinated by such books and was inspired to read them himself. Within a year he was a rapid reader and had included science in this sphere of interest.

Tom's father, Samuel Edison, Jr., was an unusual man in many respects. Although over 50 years of age at the time, he won a wrestling match in Port Huron which was open to all comers, including the soldiers at Fort Gratiot. He was known to hike to Detroit, a distance of 60 miles, in order to save carfare. He stood over 6 feet, "was straight as an arrow," and looked like an athlete when he died at the age of 92. Besides carpentry, Tom's father engaged in the lumber, grain, and feed business although he could scarcely spell well enough to write out a bill. In an attempt to supplement his income, he built a 100-foot observation tower on the corner of his property, for which he charged 25¢ for one to mount the winding stairs. A telescope at the top allowed an excellent view. The tower did not prove to be a financial bonanza although on Sundays in the summer many picnickers who came to the local park climbed it. The novelty soon wore off, and although the admission charge was gradually reduced to 5¢, the tower fell into disuse. Because of expanded activities at Fort Gratiot during the Civil War, in 1864 the Edison home was requisitioned by the Government. Edwin Stanton, President Lincoln's Secretary of War, decided the tower must go and ordered Sam Edison to "put the ax to it." The elder Edison half-heartedly tried to bring it down without success, but he apparently weakened the structure since a windstorm toppled it into the river the spring of 1865. Thus, young Thomas did not have to go beyond the family premises to see rather extraordinary innovation being practiced.

By the age of 10, Tom had accumulated some chemicals purchased at the local pharmacy, various scraps of metal he found, and electrical batteries which he had made. Using a science dictionary which described various experiments, Tom conducted some of them in his bedroom. Such things as spilled acids and muffled noises from certain poorly controlled chemical reactions resulted in parental orders to transfer all of his paraphernalia to the basement. Where other boys would be earning and saving money to buy baseball bats or skates, Tom spent what money he could accumulate on chemicals and accessory equipment, which required a total of about 200 containers. When Tom was eight, the other three Edison children were 26, 24, and 22 years. Thus, the boy had no playmates within the family, and since he did not attend school, was forced to play mostly by himself. What could be more exciting for a lonely boy than experimenting with his chemicals?

Chemicals, materials to build batteries, as well as wire and tools for fashioning telegraphic equipment were usually beyond the young experimenter's budget. Thus, at an early age Tom became a business man to get money for his scientific experiments, many of which were quite ingenious. In pursuing such ventures, he usually enlisted other boys in secondary roles.

Tom had located a discarded circular saw and used this to saw firewood to sell around town. The drudgery of turning the

5

saw by hand did not appeal to the boys, so Tom decided to build a steam engine to run it. Using an old boiler, stovepipe, and other scrap material, he set up the power saw in an old shed hidden by a high board fence. The contrivance worked fine until the steam engine blew up, throwing one of the boys through the fence and setting the shed and fence on fire. The nearby property owners hurriedly formed a bucket brigade to a town pump about a block away, and spread of the fire was averted. Since no one was seriously hurt, Tom escaped with only verbal criticism for another of his "crazy ideas" (1).

[Tom's reputation as an owner of scientific equipment did lead to a more successful episode—and one appreciated more by the townsfolk.] One winter evening the local hotelkeeper's daughter slipped and fell on the ice, striking her head. She remained in a coma for hours despite the frantic efforts of the local physician to revive her. Late in the night, Dr. Travers sent a cousin of the girl to the Edison home with the request that young Al bring an electric battery to the hotel immediately. The two found that carrying the heavy battery over the icy streets was no mean task. On arrival, the battery was placed at the girl's bedside, and on directions from the doctor, Tom and the cousin each grasped an electrode of the battery and used their free hands to massage the girl's body. After some time, at a point when Tom was gently stroking the patient's forehead, she opened her eyes and recognized her mother. As a result of this incident, the fame of Tom's battery spread, and Tom with battery was called many times "to relieve some sufferer." The money received for such services—usually a dollar a trip—was added to the coffer for subsequent purchase of chemicals and auxiliary equipment.

[At the age of 11 Tom began to raise and market vegetables. His partner was a neighbor boy named Michael Oates who, although older than Tom, had become his man Friday in carrying out experiments and the like, where more than one pair of hands was needed. A horse and cart were hired to use in selling the produce door-to-door in Port Huron and surrounding territory. Although the venture was reasonably successful financially, Tom considered such work monotonous. When a job as a train-boy became available, he persuaded his father to apply for the job for him.

In 1859 the Grand Trunk Railroad, a division of the Canadian National Railway, completed a line from Portland, Maine to Sarnia as well as a single track road from Port Huron to Detroit. The latter was called the Chicago, Detroit, and Canada Grand Trunk Junction Railway. Rail cars were ferried across the St. Clair between the terminals at Sarnia and Port Huron. Tom's new job was on the mixed passenger-freight, leaving Port Huron daily at 7 a.m. for the approximately three-hour run to Detroit, returning at 9:30 p.m. He was originally the water boy but soon graduated to newsboy and general vendor. Tom didn't have much competition for the job, probably because there was no assurance of any net income, and it is likely the 14½-hour day, seven days a week, did not appeal to many.

Tom soon had a profitable business not only selling newspapers, magazines, candy, peanuts, apples, sandwiches, and the like, to train passengers, but also using free space in the baggage car to bring various produce from Detroit to sell at retail during the long stops at the various train stations along the way. He set up a news depot at Port Huron and hired a boy to sell papers and magazines brought from Detroit. In season, he hired a second boy and operated a store selling fruits and vegetables not available in the Port Huron area.

A whole new world opened up to him as he learned about the operations of the wood-burning steam locomotive, the various machines in the railroad shops at Detroit, and the telegraph offices at the rail stops. Even more important, the long layover at Detroit allowed him several hours each day for visits to the public library. Many books on chemistry and other sciences not previously available to him were read, leading to more experiments which he wished to do. Since his long working day allowed him little time with his chemistry, he got permission from the train conductor to store various materials in one of the three compartments of the baggage car and thus be able to play around with his chemicals during the day in Detroit.

Looking for other means of earning money, Tom bought a secondhand printing press in Detroit, set it up in the baggage car, and began publishing a weekly newspaper. News items dealt primarily with people residing in the towns served by the Port Huron rail line. However, by means of his friendly contacts with the telegraph operators at the railroad stations, he was often able to publish late news which had not yet appeared in the Detroit papers. The Civil War was then in progress, and the populace was very news conscious. Publishing a paper for adults undoubtedly served as a severe test for Tom's spelling and grammar, which at that time were not the best.

The 14-year old Thomas not only had a highly remunerative job for one of his age, but he was very happy with his various side activities. His self-education was progressing rapidly. However, an accident occurred in the summer of 1862 which changed his good life. One day Tom's train passed over a "curled-up rail," causing the baggage car to lurch violently. A tall bottle containing sticks of yellow phosphorus immersed in water fell from Tom's storage rack. Exposure to air caused the phosphorus to ignite spontaneously. Tom was not present at the time, but luckily the conductor was nearby and put out the fire with little damage resulting. Tom, his chemicals, and printing press were dumped at the next station by the indignant conductor. Tom retained his job, but the chemicals and printing press thenceforth remained at the Edison home. Since Tom was handling such sophisticated chemicals as yellow phosphorus, one can conclude that his chemical knowledge was well beyond that of the chemistry-set dabbler. His father, by this time recognizing the value of his son's scientific interests, built a room and helped Tom equip a laboratory on the ground floor of the observation tower which, by this time, was attracting few paying customers.

Along with chemistry, Tom had tinkered with telegraphy in his experimental bent. He and James Clancy, another lad living on the military reservation, set up a telegraph line between their two homes, a distance of over one-half mile. On his train job Tom had an opportunity to learn a great deal about this new technological development and to observe its importance. In general, telegraph lines followed the right-of-way of the railroads. In villages and towns the telegraph office was usually part of the railroad station. The combination passenger and freight agent was likely also the telegraph operator, and he lived at the station. The telegraph was used to schedule and dispatch trains, highly important since most lines were single-track.

Tom had a good working relationship with the Detroit telegrapher and those along the Grand Trunk line, a situation which Tom encouraged by gifts of newspapers, magazines, and other merchandise. To increase the sales of his newspapers at the various stops, the newsboy would have the Detroit operator wire ahead pertinent news stories which the local operators would chalk up on the bulletin boards used to announce the arrival and departure of trains. Beginning in the autumn of 1862, Tom received several months of apprenticeship instruction from J. U. Mackenzie, the operator at the Mt. Clemens station. Tom hired another boy to take over his train job from Mt. Clemens to Port Huron, while he retained that from Mt. Clemens to Detroit. He spent the nights at the Mackenzies, taking lessons in telegraphy and assisting the train dispatcher. Tom had previously rescued Mackenzie's son Jimmy from being run over by a shunting boxcar during switching at the local station. Mackenzie knew of Tom's great interest in telegraphy and offered the apprenticeship training as a token of his gratitude, Tom paying only a nominal sum for his board while living with the Mackenzies. Tom already knew the Morse code, and to Mackenzie's surprise, the boy showed up for the first day of his lessons with a neat set of telegraph instruments which he himself had made at a Detroit shop. ⌊By early 1863 Tom was qualified for employment as a second-class or "plug" telegrapher.⌋

It has been said that electrical engineering began with the telegraph. It reached its heyday during the Civil War. The telegraph followed the warring armies, connecting field commanders with their superiors and news reporters with their papers. The Union Army had about 1,500 operators on its payroll. In his memoirs, General Grant said of them, "Nothing could be more complete than the organization and discipline of this body of brave and intelligent men." By 1861 telegraph lines had reached the West Coast. The demand for operators skyrocketed. It offered an exciting career to those seeking adventure and travel.

The telegraph was actually an extension of the ancient art of sending messages by fire or smoke, as well as the later use of the semaphore, an apparatus consisting of a number of levers, held in a series of positions to convey messages to one at a distance (2). Signalling by flags held in various positions, one in each hand of the sender, is a form of the semaphoric art. In telegraphy a message is conveyed by changing the electrical conditions at the sending end of the circuit in accordance with an agreed code and detecting these changes at the receiving end. Many contributed to the development of the telegraph, but in the United States it was Samuel Morse who developed a successful electric telegraph and a practical code based on dots and dashes. Morse first attempted to use pencil markings on a paper tape, but this soon gave way to a sounder at the receiving end, with a sharp click at the beginning and end of each signal. The dot is a mark of unit duration, the dash a signal three times as long, and the code for any letter, number, or punctuation mark, is an arrangement of dots and dashes separated by dot-length space signals.

In Edison's day all telegrams were sent and received manually. The essentials were a key (switch) and electromagnet (sounder) at each end, in series with a wire and a battery. The first commercial installation, from Washington, D.C. to Baltimore, was a single wire using the ground as a return line. Longer distances used a loop—*i.e.*, a two-wire system. Originally, primary batteries were used entirely as the source of electricity; it was not until much later that storage batteries and dynamos were used. Each telegraph office had to have its electric batteries, with main lines having from 25 to 350 cells. Thus, from the standpoint of the telegraph operator, next to the art of sending and receiving, the task of keeping the batteries in good working condition was most demanding.

Electricity was first generated for useful purposes by chemical reaction. Alessandro Volta, an Italian physicist, constructed the first primary electric battery, using a series of silver and zinc discs separated by layers of paper or cloth soaked in sodium chloride solution. When the outer discs were connected with a wire, current flowed from the zinc (anode) to the silver (cathode). Zinc dissolved, and hydrogen was evolved at the silver discs. The chemical reactions involved were:

$$Zn \rightarrow Zn^{2+} + 2\,electrons$$

$$2\,H_2O + 2\,electrons \rightarrow \overline{H}_2 + 2\,OH^-$$

Silver, being below hydrogen in the electromotive series of metals, does not react and serves as an inert electrode.

The Volta battery, or pile as it was called, was used as the source of current for making such epic developments in electricity as those by Davy and Faraday in the early part of the 19th

11

century. The Daniell cell was developed in 1836 and consisted of a zinc anode immersed in zinc sulfate and a copper cathode in saturated copper sulfate solution. The two liquids were separated by a porous diaphragm. The zinc dissolved to form zinc sulfate, and copper was deposited from the copper sulfate at the cathode. Thus, the net reaction was:

$$Zn + Cu^{2+} \rightarrow Zn^{2+} + Cu$$

The Daniell cell gave a steady voltage of about 1.08 volts at moderate current drains. This cell was widely used for the telegraph during its early days.

The stronger and longer life Grove battery was used widely for telegraphy when Edison worked as a telegraph operator. Zinc (amalgamated at the surface with mercury) in dilute sulfuric acid was separated by a porous pot from a strong nitric acid solution containing a platinum cathode. The hydrogen liberated at the cathode was oxidized to water by the nitric acid. Robert Bunsen (1811–1899), a famous German chemist, developed a similar battery but used carbon for the cathode. Either cell would give about 1.9 volts and was capable of delivering heavy currents at sustained voltages as long as the concentration of the nitric acid was maintained. The oxides of nitrogen, liberated as the nitric acid was consumed in the oxidation of the hydrogen, caused the battery room of a typical telegraph station to be filled with noxious fumes. However, in some ways they were preferable to hydrogen, liberated from the common zinc–silver–sulfuric acid or zinc–copper–sulfuric acids cells, which could result in fires or explosions.

A further advance was made when the nitric acid of the Grove-Bunsen cell was replaced by potassium or sodium dichromate in sulfuric acid. Since the dichromate did not harm the zinc anode, the diaphragm could be omitted and a single solution used. The overall chemical reaction of this cell was as follows:

$$3 \text{ Zn} + 7 \text{ H}_2\text{SO}_4 + \text{Na}_2\text{Cr}_2\text{O}_7 \rightarrow$$
$$3 \text{ ZnSO}_4 + \text{Na}_2\text{SO}_4 + \text{Cr}_2(\text{SO}_4)_3 + 7 \text{ H}_2\text{O}$$

This was the battery used by Edison in his early work on the electric light, as we shall see in Chapter 2. It had the advantage over those using nitric acid in that the hazards inherent in handling this acid-oxidant were avoided, and there were no gaseous by-products from the overall chemical reactions within the cell. Thus, there was a great deal more chemistry involved in the use of, and studies dealing with, electricity than in later years after the dynamo became the widely used source of electric current.

Although America led the world in developing and using telegraphy, she trailed the progressive European countries in basic and applied chemistry. The cradle of modern chemistry was in France where, at the end of the 18th and beginning of the 19th centuries, many great scientists such as Lavoisier and Dumas carried out their work. However, the French were slow practitioners of their chemical knowledge to industry. Until about 1860, England was the country where applied chemistry, especially inorganic chemistry, was most advanced. The extensive British textile industry was mainly responsible for the early development of their chemical industry since it required large quantities of soda ash, sodium hydroxide, bleaching materials, soap, and dyes. However, it was Germany, although late on the chemical scene, who pioneered the marriage of basic and applied chemistry which resulted in its domination of the chemical industry from about 1870 to World War I.

Some of the early industrial chemical operations were extremely primitive. In 1777 the famous George Macintosh became engaged in the manufacture of a lichen dye for wool and silk. To prepare the dye, *Lichen tartaveus,* obtained by scraping rocks in the Western Highlands of Scotland, was stirred in wooden troughs with ammonia solution. The latter was obtained from the distillation of human urine collected door-to-door in Glasgow. In 1786 the Macintosh family entered into the manufacture of ammonium chloride prepared from chimney soot and urine. It was not until 1819 that Charles Macintosh produced ammonia by the destructive distillation of coal tar.

In 1746 a factory was built in England to make sulfuric acid by burning a mixture of sulfur and saltpeter (potassium or sodium nitrate). At about the turn of the century, it was found that the oxides of nitrogen were catalytic in converting the sulfur dioxide to the trioxide if a continuous flow of air were provided. Processes were worked out for recovering and reusing the nitrogen oxides which led to the highly economical chamber process for making sulfuric acid. Soda ash was made as early as 1823 in England using the Leblanc process.

The arts of glass making, cement making, and dyeing with natural products were practiced even in colonial America. Because of the hazard and resultant high cost of transporting strong mineral acids from Europe in glass carboys, these were the first chemicals made in the United States. In 1829 Charles Lennig built extensive units at Bridesburg, Pennsylvania, and at Baltimore for manufacturing sulfuric acid. Grasselli began the manufacture of sulfuric acid and other heavy chemicals in 1839. Hydrochloric acid was prepared from common salt and nitric acid from Chilean saltpeter by treatment with sulfuric acid.

The manufacture of soda ash in the United States came much later, probably because this chemical is non-corrosive and easy to ship in bulk. In fact, soda ash was never made in the United States by the Leblanc process. In 1882 a plant was built at Syracuse. New York, to prepare it by the Solvay process. Other inorganic chemicals manufactured in the United States prior to 1870 included lime, lead pigments, chlorine bleaches, bromine, potassium cyanide, potassium dichromate, zinc oxide, acid phosphates for baking powder, and various pharmaceuticals. In 1870 the Du Pont Company, within 50 years to become the largest U.S. chemical company, was manufacturing only black powder, a mixture of saltpeter, sulfur, and charcoal (*3*).

Although most American colleges offered some instruction in chemistry, it was usually as one or more courses in the curricula for medicine, mineralogy, or even botany. Prior to the Civil War, only Amherst, Harvard, Rensselaer, and Yale had anything which might be called a chemical laboratory. In 1872 when Ira Remsen, fresh from five years of study in Germany, began teaching chemistry at Williams College, he requested space and money for a chemical laboratory. In refusing the request, the college president cautioned the young upstart: "You will please keep in mind that this is a college and not a technical school." Remsen was given the chair of chemistry at Johns Hopkins when this institution was founded in 1876, and by the 1879–1880 school year, Remsen and his assistant had combined the chemical laboratory with the classroom for training chemists. This began a new era in the teaching of chemistry in the United States.

In Germany, however, Justus von Liebig had built the first university laboratory for student instruction in the 1820's. By 1850 the Germans were not only utilizing the laboratory in teaching chemistry, but after the fifth or sixth semester, the college chemistry student began research. The German technical high schools stressed applied science. There were close ties between the educational institutions and industry. The Germans became foremost as suppliers of scientific apparatus and instruments. Excellent chemical publications became available both to the German student and the industrialist.

The organic chemical industry in the United States at the time young Edison was seeking a career in science was indeed small. The only organic chemical products manufactured were those from natural products, and as a whole, these were

15

crude materials. Turpentine was obtained as a by-product from the extensive naval stores industry in the southern states whose main products were tar and pitch for caulking wooden ships. Glycerin is a by-product in the splitting of fats to make soap, but little of it was recovered. There were extensive operations in America to make charcoal from the large resources of hard woods, but very few of the retorts were equipped to recover such by-products as methyl alcohol, acetone, and acetic acid. Ethyl alcohol, as a chemical, was available in small quantities from the fermentation of grains, as was amyl alcohol from the by-product, fusel oil. Although by 1862 the petroleum industry was furnishing kerosene to replace coal oil for lamps, its products were mixtures of hydrocarbons, largely of unknown composition.

As late as the 1879 annual meeting of the American Association for the Advancement of Science held at Saratoga Springs, New York, all of the papers presented before the chemistry subsection dealt with inorganic chemistry. In fact, Ira Remsen, chairman of this chemical group, urged in an address at this meeting that organic chemistry be taught at American colleges. He pointed out that whereas American chemistry professors felt organic chemistry was unimportant, their German counterparts were emphasizing it, and, he added, "German chemists are everywhere sought after" (4).

In 1863 the opportunities for an American lad of 16 with no formal education to obtain a position as chemist in industry were practically nil. First, he would not qualify, and secondly, there weren't any such jobs available except a few as chemical analysts. Chemical industry in America was still an art not a science. Most of the well-trained chemists were at universities. Tom with his meager formal education was ineligible to enroll as a student even if he had been financially able to do so. To receive any formal college training was so far beyond his hopes it is likely that neither he nor his parents ever considered it.

Thus, the self-educated young Edison was actually very fortunate in having "at his doorstep" in a small town area a branch of an industry, the telegraph, which was utilizing in its operations two new technologies—electricity and applied chemistry. Furthermore, the telegraphy professionals, as we would call them today, needed only to demonstrate their skill in sending and receiving messages and in keeping the equipment in working order. No diploma was required to qualify for a job. As it turned out, the telegraph companies were to become leaders in subsidizing scientific research and purchasing new developments. Thus, Tom was entering a field which was not only highly technical and undergoing rapid growth but one which was to offer opportunities to those who could innovate.

It was obvious from the start that Tom did not become a telegraph operator to make more money or to become expert at sending and receiving messages but rather to have facilities for extending his self-education in the scientific field. He gave up his lucrative job on the train where he earned several dollars a day for one where he initially made less than one dollar per day. At his first job at the Port Huron local telegraph office, housed in a local jewelry-bookstore, Tom served on a 24-hour basis, using a cot in the rear of the store for occasional sleeping. Since the volume of telegrams to be handled was small, there was time to read back issues of *Scientific American*, which were stacked in the basement, and to experiment with electrical circuits of his own. He helped himself to the proprietor's fine watchmaking tools for cutting wires, working on batteries, and so on. Thus, he was eager to accept a night operator's job at Stratford Junction, Ontario, which became available and where his salary advanced from $20 to $25 per month. Here traffic was light, but the night operator had to signal the main office every hour on the hour to show that he was fully awake and on the job. Tom rigged up a clock device to do this automatically, but the scheme was detected and he was discharged.

17

This sort of life continued for about four years between 1863 and 1867 as the youth held telegrapher jobs, usually of short duration, in cities from Canada to Memphis. His four jobs in 1864 included one at Western Union's office at the Union Station in Indianapolis, followed by one at their Cincinnati headquarters. He was usually fired from a job for allowing messages to accumulate, for neglecting incoming messages to tend to an experiment which he was running, or for changing the telegraphic equipment in attempts to improve it. He was the classic example of what every manufacturer knows: a good research man usually makes a poor operator on the production line. However, because of the shortage of operators and his willingness to work the night shift, he never had trouble finding new employment. Despite his indifference to becoming a master operator, by late 1864 he had graduated from the plug class to a full operator, which meant he could take press as well as send and receive routine messages. By early 1865 he was making $105 per month at the Cincinnati office, an excellent salary in those days. Also, happily, his deafness did not interfere with his performance as a telegraph operator.*

In these early manhood years, Edison lived a Spartan life, renting cheap rooms which he filled with technical books, electric cells, and relays, and spending most of the daylight hours reading and experimenting with electric circuits when he should have been sleeping. However, his objective of using the life of a telegraph operator as a means of self-education was paying off. Cincinnati was the largest city in the Midwest at that time and offered new opportunities in books to be read and equipment to be purchased. At his various jobs he had an opportunity to

---

* Apparently a combination of after effects from his early siege with scarlet fever and an injury to his ears when a train conductor grasped his head in helping to pull him aboard a moving train had made Tom partially deaf. Later in life he maintained that his deafness was an asset in receiving telegraphic messages since he did not hear extraneous noises which would be distracting to one of normal hearing.

see how the telegraph functioned all the way from units in country villages to those at regional headquarters. He made friends of certain fellow operators and office managers who were helpful to him later. He learned methods, real and proposed, for transmitting telegrams of secret information so that others could not intercept them. Junk heaps of the telegraph stations were scoured to recover metals and wires for constructing batteries and other devices. The young innovator usually managed to help himself to the desired acids for his batteries from those stored in carboys at the various stations. In handling news reports, he met and conversed with news reporters and Associated Press personnel, many of whom were able, educated people.

As Tom went from job to job, his mind became filled with ideas and schemes for improving the telegraph. Most of the managers were not technically qualified to discuss intelligently his revolutionary ideas, and at more than one station he became known as the "loony one." His on-the-job experiments to improve the scope and efficiency of the equipment usually resulted in dismissal. He found some outlet for his innovations by constructing unique devices for electrocuting the ever-present cockroaches, by administering unexpected electrical shocks to his co-workers when they were in the men's room, and similar forms of horseplay.

By late 1867 Tom appeared to "have had it." He considered going to Brazil with a group of ex-Confederates. Luckily, mob rioting in New Orleans delayed his departure, and after learning more about the dangers and hardships in the forbidding Amazon region, he changed his mind. Instead, he contacted a friend then working in Boston who was able to get him a job in the Western Union office there.

This move was definitely directed towards locating in an area which had become known as a leader in technology. For the first time he had access to the best scientific books, such as the works of Faraday describing his experimental research in physics and chemistry (6). [In Boston there were many shops of skilled workmen and would-be inventors making various electrical devices, such as fire and burglar alarms as well as instruments for measuring various phases of electricity. Young Edison soon had himself a corner in the shop of Charles Williams, Jr., a maker of certain telegraphic and other fine electrical instruments and who later, as an associate of Alexander Graham Bell, became the first manufacturer of telephones. Tom began to receive some financial backing to develop his ideas, and by mid-1868 he had given up his Western Union duties to devote full time to his own projects. Thus, at the age of 21, the self-made young scientist was on his own.]

During the subsequent year in Boston young Edison made several good technical contributions although none was financially successful. His first patent, U.S. 90,646, filed October 13, 1868, covered an electrographic vote recorder. Using a pocket knife, Edison made the model of the invention for the Patent Office from wood. With a fellow telegraph operator backing the invention to the extent of $100, the recorder was exhibited before a Congressional Committee in Washington and before certain state legislators in Massachusetts. Both groups turned it down on the basis that it would interfere with filibustering. The lawyer handling Edison's first patent application was Carroll D. Wright, subsequently U.S. Commissioner of Labor. He later described the young inventor as "uncouth in manner, a chewer rather than a smoker of tobacco, but full of intelligence and ideas."

During this period numerous inventors and artisans were working on various telegraphic devices such as alarm systems, call-boxes for local communication, so-called tickers for reporting the price of gold and stocks to local brokers, as well as the sending and receiving of messages by automatic methods. [Young Edison made certain improvements in the Callahan stock ticker, probably the best of such devices at that time, but the business venture attempting to use these improvements ended with little recompense for his work. At this time he also attempted to develop his ideas for a duplex telegraph—*i.e.*, sending two messages simultaneously on a single wire—and had a backing of $500 on this project. However, a field test carried out from Rochester, New York, to New York City was a failure, apparently because of a misunderstanding as to the proper operational procedure on the part of the operator at the other end of the circuit.

Although most of his Boston activities dealt with electricity, Tom was motivated to continue his chemistry studies by several factors. The *Boston Journal of Chemistry,* initiated in 1866 and one of the earliest chemically oriented periodicals in the United States, was certainly available to him. In addi-

Courtesy Edison National Historic Site,
West Orange, N. J. *(5)*

Edison, as a young telegraph operator, spent most of his free
time experimenting and studying.

21

tion, several of his operator friends were attending classes part time at Harvard, and the challenge of being able to stump these college boys with questions concerning electricity and chemistry was too delightful to resist. He apparently read of Nobel's research work on taming nitroglycerin by adsorption on kieselguhr and decided to try his hand at alternate methods. A small quantity was prepared, and portions were mixed with various inert materials. When a sample stored behind a stove exploded, the remainder of the trinitrate was hastily discarded in a storm sewer leading out to the nearby ocean.

After the duplex failure, Edison's financial credit and reputation in Boston took a definite turn for the worse. Not one to hesitate to move on to greener pastures when a bad situation developed, Tom borrowed a few dollars from one of his operator friends, as he had frequently done as a telegrapher, and took the night boat to New York. He left behind all of his possessions, including books, tools, and other equipment, as partial collateral for the many debts he had accumulated. In New York he called on Franklin L. Pope, chief engineer for the Gold Indicator Company which rendered ticker service on gold prices to about 300 brokers. The system and equipment being used had been invented by Dr. S. S. Laws, a brilliant but temperamental fellow, who had studied at the College of New Jersey (now Princeton University) under the noted Joseph Henry. Readings from the central indicator were transmitted electrically to dial instruments located in the offices of the subscribing brokers. Pope knew of Edison's work on a stock ticker in Boston and befriended him by allowing him to sleep on a cot in the Gold Indicator's battery room until he could find employment. For several days Tom lived on five-cent meals of apple dumplings and coffee and during the day studied the operation of the Laws indicator, a cumbersome clock-like apparatus.

Rivaling those of the storybook heroes created by Horatio Alger, Tom's fortunes suddenly took a turn for the better. One day, shortly after Tom's arrival, Laws' central equipment broke down, and neither Laws nor Pope could detect the trouble as errand boys from brokers' offices clamored for resumption of service. Edison fortunately discovered that a contact spring had broken and, at Laws' urgent appeal, quickly made the necessary repairs. A subsequent conference with Laws resulted in the youth's being hired as Pope's assistant; when, in July 1869, Pope resigned from his position to go into business as a consulting electrical engineer, Edison took over his job at a princely salary of $300 a month. Apparently Edison's suggestions to Laws on how to improve his apparatus had convinced the latter that here was a man worth far more than a maintenance engineer.

Although Edison jumped from the role of pauper to one in a very responsible position earning the equivalent of about $850 per month, based on 1970 dollars, he held the job only a few months. Laws' company was bought by Western Union in its drive to consolidate the telegraph and allied companies, and although he was asked by the new management to remain, Edison refused and joined Pope in the latter's new consulting venture. This is one of the many examples where Edison ignored financial security to chose an uncertain but more challenging future. Continuing his drive for self-independence, within less than a year Edison terminated his partnership with Pope in a friendly manner.

By 1870 Western Union Telegraph Company was fast becoming the giant in the telegraph industry. Included in its acquisitions was the Gold and Stock Telegraph Company which utilized the Callahan process in supplying stock quotations. General Marshall Lefferts, a former Army telegrapher, had become Western Union's Eastern Superintendent. He was familiar with the improvements Edison had made on Laws' indicator during his short employment there, and Lefferts was reported to have said at the time, "Edison is a genius and a very fiend for work." Lefferts now engaged the said fiend to improve the Callahan equipment. Edison accomplished this in about three weeks and received not only $40,000 for turning over these and certain other improvements to Western Union, but also an order for 1,200 of the improved stock tickers. Early in 1871, in a letter to his parents, he commented, "I am now what you Democrats call a Bloated Eastern Manufacturer."

The 24-year old Edison now set up operations on the top floor of a three-story structure on Ward Street in Newark, New Jersey and soon employed up to 50 men. The shop operated on a two-shift basis, with Edison as foreman on both shifts. He would nap on a cot when he felt the need for sleep and when the work permitted. Owing to his deafness, he could fall asleep even under very noisy conditions and appeared able to relax completely when sleeping. By late 1871 his operations had increased to three shops. He hired and retained excellent craftsmen and thus was able to spend a part of his own time on new developments. On Christmas Day, 1871, he was married to Mary Stilwell, a local girl. Even his marriage appeared to have little effect on his habit of long working hours.

In the early 1870's, a group of private investors purchased the rights to a patent of George D. Little and founded the Automatic Telegraph Company. The Little process utilized a moving tape having perforations corresponding to the Morse dots and dashes for sending, and the dot and dash signals were recorded on another tape at the receiving end. Morse's original sender included a comb-like arrangement that made and broke contact, and his receiver converted electrical pulses to wavy pencil lines on a paper tape moved by clockworks. This technique never worked well and gave way to sending by the hand-operated key and receiving by the clicks of the electromagnet armature. Little attempted to develop not only an automatic system but one much faster than sending by hand. His equipment proved to be defective in a number of respects, and Edison was hired by the Automatic Telegraph group to see

24

if the operation could be improved. He succeeded in developing equipment whereby over 200 words per minute could be sent as compared with 40-50 maximum by hand.

[Edison spent about five years in Newark as a manufacturer and developer of diverse products, mostly for the telegraph companies./] The Edison Universal Stock Printer, improved over the years, came into use in almost all security exchanges in America and Europe. In 1872 Edison applied for 38 patents on improvements and new models of stock tickers, automatic telegraphs, and related instruments.]

Chemistry played a major role in the successful development of automatic telegraphy. The receiver consisted of moist, chemically treated paper which passed over a roller which was touched by a metal needle connected to the sending terminal. A mark occurred when, and only when, the current was passing through the stylus and the wet paper into the grounded metal roller. Several chemical solutions were used at different times to get the color effect. The first successful chemical combination appears to have been a potassium iodide–starch solution; oxidation occurred at the point of contact, giving a blue color caused by the iodine-starch complex. Edison is said to have favored "the ferric cyanide of iron solution." This was probably ferric ferricyanide which gives a green or blue solution. Hydrogen is formed at the stylus under conditions of electrolysis, the ferric ion is reduced to ferrous, and insoluble ferrous ferricyanide is formed, a deep blue precipitate known as Turnbull's Blue. Edison also developed and employed an improved electromagnet shunt to produce a momentary reversal of current at the instant the battery was shut off, thus distinctly ending the chemical marks.

At one time the Atlantic and Pacific Telegraph Company, which had purchased the Automatic Telegraph Company, had 22 automatic stations linking the eastern seaboard cities with Buffalo, Chicago, and Omaha. However, automatic telegraphy was not successful at that time, probably because of the opposition of the operators. Telegraph operators were one of the first groups of American workers to unionize. Also, Jay Gould used automatic telegraphy as one of the tools to gain control of Western Union. Once consolidation was achieved, there was less incentive to spend the necessary capital to convert manual stations to automatic stations.

Early in 1873 Edison undertook further work on his duplex system of sending telegraph messages, a project which had interested him ever since his days as an operator. Western Union was interested, and he was allowed to use facilities at their New York station at night when the lines were free. Before long he developed the quadruplex, a technique for sending four messages simultaneously over one wire, two in each direction. It operated successfully and was a major technical triumph for Edison. Its use greatly cut down on the need for new lines to handle the ever-increasing volume of telegraphic communication; after it was fully installed on Western Union lines in 1876, the company reported an annual saving of $500,000. The quadruplex was widely used for about 30 years.

In the present age of telephone, radio, and television, it is easy to forget that the telegraph was the sole means of communication between distances in the 1870's. In spite of the basic importance of Edison's contributions to automatic telegraphy and his invention of the quadruplex, he received only little more than promises for the former and $30,000 for the latter. In 1875 his quadruplex invention was the subject of litigation between Western Union and Atlantic and Pacific Telegraph. In this bitter court battle, the lawyers of Western Union described Edison as a rogue inventor doing business with Atlantic and Pacific when he was under contract to them. Gould's lawyers described him as a simpleton, who had allowed Western Union to place one of their own men, G. B. Prescott, on Edison's basic quadruplex patent as a co-inventor and thus automatically gain control of 50% of the patent rights. In a few years the two companies merged, with Jay Gould gaining control and operating the monopoly under the name of the larger and older of the two companies, Western Union Telegraph Company. Gould dropped the development of automatic telegraphy and stopped payments to the stockholders of the Automatic Telegraph Company and to Edison for his work on that development. Thus, for his major contributions in the telegraph field, Edison received hardly enough to cover expenses, and, as he said later, he got "nothing for three years of hard work." The manipulations of the financiers which led to the amalgamated telegraph system and its effect on Edison are described in the excellent biography by Matthew Josephson (7).

While working on the automatic telegraph, Edison had studied paraffin-coated paper to use as the perforated tape. Finding it unsuited for this purpose he introduced it as a wrapping for candies and the like and is considered to be its developer.

Following up on discoveries relative to the electrical conductivity of compressed carbon particles, Edison devised an instrument for measuring small fluctuations in temperatures which he called the tasimeter. A hard substance having a high degree of thermal expansion is clamped firmly at one end, with the other end resting against a carbon button. An electric circuit is connected between the molded carbon and the metal plate holding it, which also includes a battery and a sensitive galvanometer. The small end of a cone-like funnel is directed to the clamped specimen, such as a hard rubber vulcanizate, and the other end is directed at the object whose temperature change is to be measured. The heat of a hand held 30 feet away could be so detected when the vulcanizate and cone were at a temperature around 75°F. Edison thought that the tasimeter could possibly be used to detect the approach of a ship to an iceberg. However, there are apparently no records of any such testing. In fact, it is not known if any scientific use was ever made of this instrument, a probable forerunner of present-day radiation thermometry. It shows, however, that Edison did spend time and money on devices which were not directed towards any possible profit-producing venture.

Edison's lack of adequate compensation by the business moguls, who used him largely as a pawn in their struggles to control the telegraph empire, was typical of the fate of most inventors in those days. Charles Goodyear, who discovered a successful process for vulcanizing rubber in 1839, died heavily in debt in 1860. Samuel Morse received only a pittance from his invention and demonstration of practical telegraphy. Alexander Bell, inventor of the telephone, found himself on the verge of ruin in 1875 and wrote to a friend, "The cares and anxieties of being an inventor seem more than flesh and blood can stand." However, Edison learned the facts of business life when he was still in his twenties. Apparently from this experience he resolved to place himself in a position where he would have greater control over the new developments which came from his research. Toward that end he gave up most of his manufacturing operations in 1876 and moved to Menlo Park, New Jersey, where he set up operations in facilities which he himself owned. With utmost faith in his own abilities, he characteristically burned his bridges behind him and started a new life where research and development, not manufacturing, would predominate.

Chapter 2 Chemistry of the Light Bulb

ONE GROUP WHICH HAD FAITH in Edison's approach to electric lighting was his staff. During the approximate one year when approach after approach was tried and abandoned, his employees remained loyal even though he was not always able to meet his payroll on time. Most were older than he, but none questioned his leadership. The total number of employees at the Menlo Park Laboratory prior to the successful development of the light in October 1879 was only about 30 people. Probably not more than one-third of these were working directly or indi-. rectly on the incandescent light.

Every American is a direct beneficiary of Edison's development of the incandescent light and subdivision of the electric current. Electricity is now brought safely and economically into the most humble home, not only for lighting but also for operating numerous labor saving and entertainment devices, for cooking, and even for heating. With apologies to the late Winston Churchill who paid homage to the fighter pilots who won the battle of Britain with his tribute, "Never in the field of human conflict was so much owed by so many to so few," one can also truthfully say of Edison and his associates that never in the field of research and development have so few in such a short time done so much for so many.

Pine knots and other burning stick-torches were the first sources of lighting both for the traveler and for use in buildings.

Historically, man had noticed that cooking fat which fell into the fire acted as fuel for lighting. This led to lamps which used olive and nut oils, fish oils and animal greases, and whale oil in hollow stones and shells, with moss used as wicks. Pottery lamps were developed along the Mediterranean, and later metal lamps were made in Europe which could be tipped to adjust the flow of the grease. The American colonists brought over iron Betty lamps which had been in use for generations. This Betty lamp had a cover with a wick down through a metal tube.

Although candles have been used from the time of Christ, they became widely used during the period 1700–1850, especially for those who could afford more gracious living. Combinations of tallow fat, beeswax, and spermaceti, obtained from the sperm whale, were used. All of these are esters of aliphatic alcohols and aliphatic acids and burn with clear flames because of the relatively high ratios of hydrogen to carbon in the molecules. Such candles were a pleasant change from the sputtering grease lamp or smoking pine knots. One problem of the candles made from animal fats was that they had to be kept in a metal box when not being used so they would not be eaten by mice.

The oil lamp continued, however, to provide most of the lighting in America. Whale oil was the primary fuel until about 1830 when the whale population off the New England coast became depleted as a result of the large whaling fleet operating out of northeast ports. An oil called camphene derived from the redistillation of turpentine was the dominant lamp illuminant in the 1840's. However, it tended to be smoky, so mixtures with alcohol were also marketed to obtain improved burning characteristics. Use of the volatile alcohol led to a number of explosions because of explosive mixtures of fuel and air in the chimneys when the lamps were lighted. A decade later, coal oil, obtained by heating cannel coal in the absence of air at temperatures in the 950°–1400°F range, largely replaced camphene.

As every student of chemistry knows, the liquid and tarry components in coal are largely aromatic. That is, the ratio of carbon to hydrogen is high—*e.g.*, C_6H_6 for the simplest aromatic hydrocarbon, benzene *vs.* C_2H_6 for a typical aliphatic. Such hydrocarbons burn with smoky, sooty flames. However, cannel coal, which is believed to be derived from the seeds and spores of plants rather than the wood, bark, and leaves, burns with a bright, non-smoky flame and is the favorite coal for use in fireplaces if an aesthetic effect is desired. Thus, oil derived from cannel coal is high in aliphatic and naphthene hydrocarbons, which have typical chemical formulas of C_nH_{2n+2} and C_nH_{2n} and burn without smoke in properly adjusted lamps. Coal oil paved the way for kerosene, which appeared on the market soon after E. L. Drake began operating the first successful oil well in 1859 at Titusville, a village in northwestern Pennsylvania. Kerosene, often erroneously called coal oil, soon became the universal illuminant used in lamps and had the largest use of any petroleum product until the turn of the century. Oil-burning lamps with cylindrical adjustable wicks, a means of controlling the air flow, and glass chimneys became available at low cost. These, plus an ample supply of kerosene, made possible reasonably good lighting in the homes of rural America.

Hydrocarbon gases, derived from the high temperature gasification of coal, were first made commercially and used by a factory located in Birmingham, England, in 1802. Ten years later, the streets of London became the first in the world to be lighted by gas. By 1830 Baltimore, Boston, and New York had installed gasworks, but their output was confined almost entirely to lighting streets and public buildings. So-called water gas, a mixture of carbon monoxide and hydrogen, was developed in 1834 by A. F. Selligue of France by the reaction of steam with incandescent coke. By 1860 manufactured gas was available in 400 urban communities in the United States and became widely used for lighting streets, homes, factories, and offices. So effective was street lighting by this new invention that some of the clergy accused city fathers of blasphemy because of their attempts to turn night into day. Hotels using gas lighting in their rooms found it necessary to put up signs instructing their guests not to blow out the flames but rather extinguish them by turning the valves. By the late 1870's, New York City had several gas companies, all doing a thriving and profitable business.

Even before Michael Faraday discovered the principle of the dynamo in 1831, it was known that lighting could be produced by electricity. Early in the 19th century, Sir Humphry Davy had produced an electric arc using a battery of 2,000 cells. He attached a piece of charcoal to the ends of each of two battery wires, brought them in contact, and then separated them. At once the intervening space was filled with flame. However, nearly three quarters of a century passed before lighting by means of the electric arc became a practical reality. In the interim applied electricity had been limited to telegraphy and, in a small way, to electroplating. Electrical engineers were few, as indicated by the fact that the American Institute of Electrical Engineers was not founded until 1884.

Edison worked four years as a telegraph operator and from 1869 to 1876 had engaged in the invention and manufacture of various telegraph systems and equipment. In 1876 he invented the electric pen, which was the first industrial product to utilize an electric motor for power. The "pen" consisted of a needle on the end of a plunger which was used to make perforations at the rate of about 8,000 per minute. The miniature electric motor at the top of the pen drove the needle by means of cams on the rotating motor shaft. The perforations were usually made over written matter and then used as a stencil for making multiple copies. In 1877 he developed the carbon telephone transmitter, again an invention which involved the use of electricity Thus, it is not surprising that his friends, including Professor George F. Barker of the University of Pennsylvania, Professor Charles F. Chandler, an eminent chemist of Columbia University, and Grosvenor P. Lowrey, chief counsel of Western Union Telegraph Company, urged Edison to enter the field of electric lighting.

The arc light was the first successful means of adapting electricity for lighting. It consisted of two carbon electrodes, which required a hand or automatic mechanism to keep them the proper distance apart as they burned. As a rule, the carbons had to be replaced daily. The light fluctuated, hissed, and often threw off hot embers. It was intensely bright. In 1862 an arc light was installed and successfully used in a lighthouse on the coast of England. There was one light and one dynamo; the current was 10 amperes at 45 volts. Leaders of arc lighting in the United States were Elihu Thompson, Edward Weston, and Charles Brush. Brush installed arc lighting in the public square of Cleveland in 1878, and similar street lighting installations followed shortly thereafter. Although light from the arc lamp is derived mainly from the glowing carbon at the points of arcing, it is similar to gas, kerosene, pine knots, and other burning fuels in that oxidation occurs with the carbon being converted to gaseous products.

Students of electric lighting were soon aware that when current traversed a conductor of high melting point and high resistance, a glow or incandescence was obtained. Attempts were made as early as 1820 to make an incandescent lamp with a platinum "burner." Up to the time of Edison's successful development in 1879 literally scores of investigators developed such lamps, but none was practical. Joseph (later Sir Joseph) Swan of England devised a lamp using a strip of carbonized paper as a burner in an evacuated glass bulb. However, it was of low resistance, of short life, and Swan never solved the problems of coupling the carbonized paper to the source of current.

In 1872 the Russian, Lodyguine, developed an incandescent lamp using a graphite rod in a nitrogen atmosphere. Two hundred such lamps were installed in St. Petersburg, Russia. Again, these lamps had short lives and were unreliable in operation. Because of this widespread activity, Edison wrote to a friend in the spring of 1878 that he did not intend to take up electric lighting research because "so many others are working in that field."

However, as early as 1876 Edison had already carried out some experiments on the amount of current required to bring various non-conductors to incandescence. In September 1878 he visited William Wallace, a manufacturer of dynamos, in Ansonia, Connecticut. Wallace, in cooperation with Moses Farmer, was also in the process of bringing out an arc light system. For the first time Edison became excited with the idea of working on electric lights. He gave Wallace an order for a generator to serve as a source of current for his experiments.

Edison's basic approach was to develop an electric light which could be substituted for gas for lighting homes, offices, public buildings, factories, and the like. This meant small lights which could be individually controlled. It also meant that it would be necessary to subdivide the current from a dynamo or dynamos into separate channels much as one does in the distribution of gas. Edison, of course, was not the first to visualize this, but he was the first to conceive how it could be done successfully. Although the relationship of $E = RI$ had been postulated by Ohm in 1827, it was not accepted by many (1). More important, the energy or horsepower used by a given light was not realized to be EI. Thus, where others had tried to feed current to a number of lights from one dynamo by using high amperage, Edison visualized the use of as high a voltage as could be handled safely by the customer, thus greatly reducing the amount of current necessary to give an adequate light. Then he could use parallel wiring of individual lights rather than series (2).

This meant that lamps of high resistance would have to be used, and a source of current of constant pressure—*i.e.,* voltage—would be required.

Although there were many problems to be solved before electricity could be produced and transported to hundreds of customers from a central source, that of developing the lamp was, of course, No. 1. Edison had a very definite picture of what the requirements of such a lamp had to be:

(a) To obtain the desired resistance of 100 ohms or more, the lighting element would have to be a poor conductor of threadlike thickness.

(b) The element would have to be of very high melting point and stable up to a yellow heat (*3*).

(c) Since most solid elements combine with oxygen at elevated temperatures, the container for the incandescent light would likely have to contain a permanent vacuum or an inert gas.

(d) To be economical, the light must last hundreds of hours, with the filament being cycled scores of times from room temperature to a temperature of incandescence.

All of these problems dealt with the chemical properties of materials and physical chemical phenomena. The stability of materials at about 1000°C and higher was largely unknown. The electric furnace was yet to be invented, so few materials had been heated above the temperature of the muffle or blast furnace. Skeptics, many of whom were university scientists, pointed out (as information on Edison's approach to the electric light became public) that carbon would very likely break down at incandescent temperatures. They predicted further that if by chance such a carbon breakdown did not occur quickly, it would shortly burn anyway since a long term vacuum could not be maintained in an electric light bulb, and if, by the grace of God, a successful light should be developed, there wouldn't be enough copper in the world to bring the electric current to customers now receiving gas for lighting.

35

Courtesy Edison National Historic Site, West Orange, N. J.

Artist's concept of winter scene of Edison's Menlo Park Laboratories. The business office-library is in the foreground. The experimental electric railroad is on the right.

Edison now had excellent facilities for carrying out applied chemical research in contrast to his contemporaries. In the spring of 1876 he had given up his small laboratory in Newark, New Jersey, turned over his manufacturing operations there to others, and moved to his new quarters at Menlo Park, New Jersey. Menlo Park was certainly not a park—in fact, not even a village. It was a flag stop, located about 25 miles southwest of New York City, on the Pennsylvania Railroad running from Jersey City to Philadelphia. The hamlet had no church or school. Most of the men who subsequently worked for Edison lived in nearby Metuchen and Rahway or on neighboring farms.

Edison had engaged his father, then living in Port Huron, Michigan, to construct the laboratory buildings. The main laboratory structure was of two stories, 25 feet wide and 100 feet long. One end of the first floor was used as a machine shop. Several laboratory furnaces were located nearby. The main test table was on the first floor which stood on two large pillars of brick extending deep into the ground. Thus it was recognized that such a support was necessary for sensitive instruments subject to vibrations. A small laboratory, used primarily for analytical and assay work, was partitioned off, as were a small library and a business office.

A wide stairway led to the second floor which housed several long tables where the bulk of the experimental work was done. Shelves covered most of the wall space on the second floor, and on these were approximately 1400 glass bottles and jars containing chemicals, oils, metals in powder and wire form, and natural and animal products of all kinds. One newspaper man was reported to have remarked that "the place looked like a drug store."

The initial construction also included three small buildings: a carpenter shop, one which later became the glass-blowing shop, and one that was used to make lampblack. In 1879 after work was expanded on the electric light, a separate brick struc-

THOMAS EDISON, CHEMIST

Courtesy Edison National Historic Site, West Orange, N. J.

Scene taken on upper floor of main laboratory building. Edison, with skull cap, is shown in the center area. Organ, in rear, was a gift from an admirer.

38

ture was built to house the machine shop, and at one end there was a boiler, a steam engine, and space for the installation of several dynamos. At the same time, a two-story brick building was erected for the business office, a drafting room, and the library. The library on the second floor of this rather elegant building had expensive furniture and was the brainchild of Grosvenor Lowrey, then president of the Edison Electric Light Company. Lowrey correctly foresaw that Edison would be having important visitors as the work progressed and insisted that Edison dress and receive these dignitaries in "proper style."

Edison's home was on a plot of land 300 x 150 feet, located about 800 feet from the laboratory. The Edisons had a barn, horses, and carriages. Water was supplied from a well whose pump was driven by a windmill. There were six other houses in Menlo Park, and these were largely occupied by Edison's key personnel or maintained as boarding houses for Edison's bachelor employees. The largest of these houses was operated by Mrs. Sarah Jordan, a distant relative of the first Mrs. Edison. This house was dismantled and reconstructed at the site of the restoration of the Menlo Park laboratory and auxiliary buildings in Greenfield Village in Dearborn, Michigan (4).

Edison's installation at Menlo Park was no doubt the first industrial research laboratory in the United States, and likely in the world, based on our present-day definition of what constitutes such an organization. In contrast to the lone inventor working in his attic or basement or industry in carrying out occasional experimental work in one corner of a factory, Edison's installation was devoted entirely to developing new things. His organization consisted of knowledgeable professional people and assistants with adequate laboratory facilities, excellent equipment and test instruments, and a broad inventory of materials. This experimental effort was backed up with expert mechanics and machinists, a glassblower, a staffed library, a business and clerical staff, draftsmen, and a patent department. Charles F. Kettering, inventor of the automotive self-starter and a noted authority in the field of invention, once said, "To my mind, Edison's greatest 'invention' was organized research." It is also fair to say that this was not only the first industrial research laboratory but also the first with a rural country club-type location.

After moving to Menlo Park, Edison slowly built up his staff and by the winter of 1878–79 employed about 30 people. An Englishman, Charles Batchelor, a watchmaker of wide experience with fine machinery, continued to be Edison's right-hand man. They had a dozen or so laboratory assistants, and based on present practice, the ratio of laboratory assistants to professionals was very high. John Kruesi, a native of Switzerland, headed the shop operations. As glassblower, Edison engaged Ludwig Boehm, who had learned his trade under Dr. Heinrich Geissler, a German manufacturer of special scientific equipment who was considered the master glassblower of his day.

Dr. Alfred Haid, a chemist, occupied the small laboratory on the first floor. Dr. Otto Moses was in charge of the library and responsible for searches in foreign publications. His excellent command of German and French made him a good contact with the foreign savants who visited Menlo Park. Edison's most trusted technical man was Francis R. Upton, a mathematician and physicist. Upton had graduated from the College of New Jersey (now Princeton University) and had studied an additional year under Hermann von Helmholtz, a noted German physicist. Upton was largely a desk man, being assigned numerous jobs involving calculations and design, tasks which Edison abhorred.

Francis Jehl, a youth of 18 who had studied chemistry at night at Cooper Union, came to Menlo Park early in 1879. He had previously worked in the office of Grosvenor Lowrey, where one of his assignments was to make multiple copies of various written matter by means of the Edison electric pen. This operation included taking care of the Bunsen cells which furnished current for the pen. Jehl's first job on being hired by Edison was to clean the 50 odd Bunsen cells on the various tables on the second floor of the laboratory. The electrodes consisted of carbon in potassium dichromate solution and zinc in dilute sulfuric acid. After each cell was cleaned in the laboratory sink, fresh electrolyte solutions were made up and used. The zinc plate was rubbed with mercury and rinsed. All binding posts were tightened, and each cell was tested before the job was considered done. After Jehl had finished his first assignment, Edison said, "I see you know the ropes." As a result of the good first impression, Jehl was retained by Edison to work closely with him during the laboratory stage of development of the electric light. Jehl was unmarried and by living at the nearby Jordan Boarding House apparently was at the laboratory at night as well as by day when there was anything to be done. In 1882 he went to Europe to introduce the Edison lighting system in various countries. After World War I he returned to the States and was subsequently engaged by Henry Ford to supervise the reconstruction of the Menlo Park laboratory complex at Greenfield Village. Jehl subsequently wrote a three-volume treatise entitled "Menlo Park Reminiscences" which details the technical developments and personal relationships which took place in this highly successful invention factory (5).

From the standpoint of financing, the Menlo Park venture was a combination of a private small business, consulting, and corporate support. Except what was needed for family expenses, Edison continued to put all of his personal income from manufacturing and royalties into the effort. His income was probably about $50,000 per year, and he alone bore the cost of the site and that of constructing and equipping the original buildings. Starting in the spring of 1876 he was paid $500 per month by

Period	Group Ia	Group IIa	Group IIIa	Group IVa	Group Va	Group VIa	Group VIIa	Group VIII			Group Ib	Group IIb	Group IIIb	Group IVb	Group Vb	Group VIb	Group VIIb	Group O
1	1 H																1 H	2 He
2	3 Li	4 Be											5 B	6 C	7 N	8 O	9 F	10 Ne
3	11 Na	12 Mg											13 Al	14 Si	15 P	16 S	17 Cl	18 Ar
4	19 K	20 Ca	21 Sc	22 Ti	23 V	24 Cr	25 Mn	26 Fe	27 Co	28 Ni	29 Cu	30 Zn	31 Ga	32 Ge	33 As	34 Se	35 Br	36 Kr
5	37 Rb	38 Sr	39 Y	40 Zr	41 Nb	42 Mo	43 Tc	44 Ru	45 Rh	46 Pd	47 Ag	48 Cd	49 In	50 Sn	51 Sb	52 Te	53 I	54 Xe
6	55 Ca	56 Ba	57 La [a]	72 Hf	73 Ta	74 W	75 Re	76 Os	77 Ir	78 Pt	79 Au	80 Hg	81 Tl	82 Pb	83 Bi	84 Po	85 At	86 Rn
7	87 Fr	88 Ra	89 Ac [b]															

[a] Lanthanide series: Nos. 58-71.
[b] Actinide series: Nos. 90-103.

Western Union to carry out research on the "speaking telegraph." In the fall of 1878 the Edison Electric Light Company was formed to support Edison's work on the electric light. This was a noteworthy event in the annals of American industry. For the first time money was being invested in faith that a research organization would accomplish its objective, and patents on inventions, yet to be made, would allow the stockholders to recover their money and at a profit.

When Edison returned in earnest to the search for a "burner" for incandescent lighting, he knew that in order to obtain the desired high resistance, a very small diameter material would have to be used. He logically turned to the high melting metals which could be drawn into fine wire, or filaments as Edison now chose to call them. Platinum had been used by others as short thick wire or rods, and its high melting point and ductility were well known. The electrical resistance of platinum is about seven times greater than that of copper, and this factor was also favorable. Dmitri Ivanovich Mendeleyev, as well as Lothar Meyer independently, had developed the periodic law of the elements in 1869, which showed that the properties of the elements are a periodic function of their atomic weights (6). No doubt Edison expected that this recent development would be of considerable help to him in choosing the most likely metals for testing. Although most of the common metals had been isolated by 1878, he would need to purify many of them, using the services of Dr. Haid.

All members of the platinum family of metals had been discovered by 1845, and Mendeleyev had arranged them in Group VIII of the Periodic Table (p. 42). This group is included among the transition elements in the center of the long periods five and six; that is, they do not fit into the groups and subgroups I through VII but are intermediate in properties between certain elements in Group VII and others in Group I that follow them in the same period. The six members of the platinum group of metals are the first six cited in Table I (pp. 50-51). The first triad—ruthenium, rhodium, and palladium—precede silver in the fifth Period, and the second triad, which are of considerably higher atomic weight, precede gold in the sixth Period. Although the three metals making up each of the two triads are quite similar in atomic weights and densities, from the standpoint of chemical and physical behavior there is also a subgroup relationship. For example, palladium, the third member of the first triad, resembles most closely platinum, the third member of the second triad. This may be noted from the melting points and resistivity values cited in Table I.

Most of the scientific articles on platinum in the 19th century were in French or German journals and thus available to few Americans. In 1879 Edison published an article entitled "Action of Aqua Regia on Platinum" in the *Scientific American* (7). There is very likely little of an original nature in this publication since Scheffer of Sweden discovered and described this reaction in 1752. However, the article does indicate that Edison was probably trying to prepare high purity platinum for his experiments.

By 1878 platinum had found wide use in analytical laboratories for crucibles. Because of its high melting point and resistance to most chemicals, such a crucible was, and continues to be, used for ignitions, fusions, and the like, where glass cannot be used. Platinum anodes in strong nitric acid were used in the Grove primary battery, an electric cell used by Faraday in his basic studies and which also found use in the early days of the telegraph. When Edison was working as a telegraph operator in Stratford Junction, Ontario, he learned that several old Grove batteries had been discarded and were in several boxes in the freight house at Goderich, and he was given permission to recover what he wished from the batteries. The agent in charge apparently thought the shiny metal anodes were made of tin. Edison recovered the platinum and added it to his private chemical supplies. Since much later he had some of this platinum in the chemical stockroom of his West Orange Laboratories, it is very likely that part of it was used in his light studies at Menlo Park. Thus, Edison's knowledge of chemistry paid off in a practical manner even as a telegraph operator.

In Edison's early work on testing filaments, electricity was supplied by a combination of cells with a rheostat in the circuit to control the amount of current. The electric generator which Edison purchased from William Wallace replaced the batteries as source of current. Such generators soon were called dynamo-electric machines, which was quickly shortened to just dynamos. Edison had suggested that the dynamo be named Faradic Machines in honor of its originator, Michael Faraday, but this apparently didn't catch on (8). Wallace's dynamo at Menlo Park was replaced early in 1879 by one of the "Long Waisted Mary Ann" type developed by Edison. This type of dynamo was not only more efficient in converting mechanical energy to electrical energy than any previously built, but it was also designed to give a constant voltage. A maximum of 110 volts was chosen since this was considered safe to bring into homes and still high enough so that the consumption of a current of one ampere, then called weber, would provide a good light.

For testing, the platinum wire was attached to heavy copper wires leading to the source of current. Wires as low as 0.001 inch in diameter were tested. Testing was first carried out in air since it was known that platinum is not oxidized by air even at glowing temperatures. Soon testing was carried out in a bell jar which allowed testing in various gases and *in vacuo*. Nitrogen was the most inert gas available since argon, helium, and the other rare gases were not yet known. It was not until 1894 that Lord Rayleigh and W. Ramsay announced the discovery of argon in the earth's atmosphere. Besides nitrogen, the effect of other gases, such as chlorine and hydrogen, on the incandescent platinum was studied. Although flammable, such gases will not burn or explode in the absence of oxygen.

However, filaments of longer life resulted when vacuum was used as compared with any gas tested. Edison observed that after platinum wire had been preheated, especially in a vacuum, the incandescent platinum was not only more stable, but it had become more brittle and could be heated to a higher temperature before melting. He then studied the intermittent heating and cooling of platinum wires *in vacuo*. This resulted in the preparation of improved filaments. The conventional vacuum pump was the plunger type with valves. To study the effect of a more perfect vacuum in preparing and using incandescent filaments, Edison tried to get a Sprengel pump. In this German-made pump, as mercury flows down one leg of a long "Y" tube, a vacuum is created in the other leg which is connected to the tube (or bulb) to be evacuated. Fortunately, Edison learned that such a pump had recently been received at the College of New Jersey and had not as yet been unpacked. Francis Upton, a graduate of that college and an employee of Edison, was able to borrow it. Its use resulted in platinum filaments which lasted for several hours before failing.

At the 28th meeting of the American Association for the Advancement of Science (AAAS) held at Saratoga Springs, New York, in August 1879, Edison presented a paper entitled "On the Phenomena of Heating Metals in Vacuo by Means of an Electric Current" (*9*). Historically, papers dealing with mathematics, physics, and chemistry were grouped together in Section A. Edison's paper came under this section since instruction in electricity at that time was conducted only in physics departments of American colleges.

In his paper before the AAAS, Edison brought out several interesting facts. One phenomenon which he pointed out was that the temperature reached by an object being heated by a flame not only depends upon the temperature of the flame but also on the mass of the object and the area of its radiating surface. For example, although others had contended that it took the temperature of a hydrogen–oxygen flame to melt platinum, he had been able to melt a platinum wire of 0.001 inch diameter by an ordinary Bunsen burner. When a similar wire but of 0.004 inch diameter was used, it could not be melted by the Bunsen burner. This simple experiment demonstrated one of the principles which Edison hoped to apply in his electric light —that is, a low radiating surface allows a high temperature, and thus a brighter light, with a minimum expenditure of energy.

Edison reported another set of experiments as follows. Platinum wire was coated with magnesium acetate and then brought to incandescence by an electric current which resulted in a coherent coating of magnesium oxide. When this coated wire was heated electrically under a "glass shade," a coating of magnesium oxide formed on the inner glass surface. When this experiment was repeated in a bell jar evacuated by a hand vacuum pump, it took much longer for the deposit to form. When the platinum wire coated with magnesia was heated in a glass bulb exhausted with the Sprengel pump, no deposit formed even when the "burner" was brought to a "dazzling incandescence." Therefore, the deposit was not formed because the magnesium oxide volatilized but rather because of a "washing action of air" on the fragile oxide by convection currents rising from the heated filament. Apparently observation of this phenomenon firmly convinced Edison that a successful incandescent light required a filament sealed within an excellent vacuum.

Dealing more specifically with the subject of his paper, Edison pointed out that microscopic examination of platinum filaments which had failed revealed deep surface cracks. Also, it had been observed that when a bulb containing a cold platinum filament was evacuated and the filament electrically heated, the vacuum within the bulb fell drastically. It was concluded that air was contained within pores of the wire and that cracking occurred if this air was expelled rapidly. A series of experiments then showed that if a platinum wire was heated rapidly to its melting point, a maximum light intensity of five candles, as measured by a photometer, was obtained. When a similar platinum wire was preheated slowly under a good vacuum and then tested in a like manner, a light of 25 to 30 candles resulted. The wire annealed by the slow heating process showed no surface cracks after failure. Other metals, including even iron and aluminum, tended to be improved by such a preheating treatment but not to the degree shown by platinum and platinum–iridium alloys. Thus Edison concluded that annealing of metals by heat to render them more pliable may be the result of the formation of surface cracks by the rapid expulsion of dissolved gases.

We now know that platinum and the platinum group of metals in general are strong adsorbents for certain gases. This characteristic makes platinum a suitable catalyst, and it is so used commercially in a finely divided form, usually called platinum black. When exposed to air, platinum primarily adsorbs oxygen, and because of the inertness of the metal, heating serves to expel the gas rather than to oxidize the platinum. Thus, Edison appears to be one of the first—if not the first—to observe the strong adsorptive power of platinum for certain gases.

Edison proceeded to tackle problems relating to the use of a highly evacuated glass bulb. The inlet and outlet wires would have to be sealed tightly within the glass, and metal and glass of near-like degrees of thermal expansion-contraction would be required. Platinum wires (14 mils in diameter) and a white lead glass obtained from Eimer and Amend Company met this requirement and were used. It is now known that platinum has the lowest thermal coefficient of expansion of any simple metal. Edison's original idea was that the glass base containing the incandescent unit would fit into the light bulb by means of a ground glass joint. Thus, when the lamp failed, it would be returned to the central lighting company, and the filament would be replaced. The decision to use a permanent high vacuum lamp meant that the glass bulb had to be sealed into one piece. In that way, once the filament failed, the light would be discarded. This was a daring decision to make since the ultimate cost of manufacturing a successful lamp was unknown.

In spite of the improved vacuum obtained with the Sprengel pump, which was as low as one-millionth of an atmosphere, Edison found that platinum filament lights could not be made which would "burn" more than a few hours. Other metals were considered, and several were tested. Of the platinum metal family, only platinum and palladium were known to be sufficiently ductile to allow drawing to wires. As Table I shows, palladium has a melting point well below that of platinum. The high melting point of osmium made it potentially attractive, but it is so hard and brittle that even today it cannot be drawn to wires. Osmium filaments for incandescent lights using a unique process were developed in Europe in 1896, but they proved too brittle to be practical. Platinum alloys of iridium containing up to about 15% of the latter can be worked and are stronger than platinum at high temperatures. Such alloys were apparently tested extensively by Edison without outstanding results.

Since a highly evacuated bulb was now being used, it was no longer necessary to restrict the study to metals of high resistance to oxidation. Some work was done with nickel, but it, as well as other metals capable of being drawn to wires, melts lower than platinum. The three metals of the subgroup of Group VI cited in Table I have high melting points. In fact, tungsten is the highest melting metal. Chromium had been isolated in 1798, but it was only a laboratory curiosity in 1880. Edison apparently worked with it unsuccessfully. It is now known that chromium must be extremely pure to have substantial tensile ductility. Although molybdenum and tungsten had been

TABLE I.   HIGH

Name	Chemical Symbol	Periodic Group	Atomic Weight
Ruthenium	Ru	VIII	101.07
Rhodium	Rh	VIII	102.91
Palladium	Pd	VIII	106.4
Osmium	Os	VIII	190.2
Iridium	Ir	VIII	192.2
Platinum	Pt	VIII	195.09
Copper	Cu	I	63.57
Silver	Ag	I	107.87
Gold	Au	I	196.97
Iron	Fe	VIII	55.85
Cobalt	Co	VIII	58.93
Nickel	Ni	VIII	58.69
Chromium	Cr	VI	52.0
Molybdenum	Mo	VI	95.95
Tungsten	W	VI	183.85

a Measured at 0°C.

known since 1782 and 1783, respectively, both were too hard and brittle to be worked. Neither molybdenum nor tungsten could be fabricated until the epic work of William D. Coolidge and Colin G. Fink of the General Electric Company in 1906–1910 resulted in a means for making these metals ductile.

Since platinum must be near its melting point for effective luminescence, Edison worked on various devices for momentarily shutting off the current if the filament temperature became too high. One was an expansion device with a lever for opening and closing the circuit. Another was a diaphragm arrangement for the same purpose. Work was also done on coating the platinum wire with a thick layer of refractory material which served to hold the platinum filament in proper shape even if semiliquid. Zirconium oxide was probably the best coating tried and had the advantage that it becomes a fair

MELTING METALS

Density grams/cc	Melting Point, °C	Electrical Resistivity, Micro-Ohms, 20°C
12.45	2,310	7.2
12.41	1,960	4.5
12.02	1,552	10.4
22.6	3,045	9.5
22.65	2,443	5.1
21.45	1,769	10.6
8.94	1,083	
10.49	961	
19.3	1,063	2.1 [a]
7.87	1,532 ± 5	9.71
8.85	1,493	
8.91	1,453	6.84
7.19	1,875	12.9
10.22	2,610	51.7
19.3	3,410	

electrical conductor at incandescent temperatures. An attempt was also made to construct a device whereby the light of the incandescent filament would be reflected and condensed by a concave reflector onto a refractory material, such as zircon, $ZrSiO_4$.

None of these modifications proved practical, and by September 1879 Edison had largely abandoned platinum in his work on filaments. Lights "burning" only a few hours had been developed, and they were inefficient since relatively low incandescent temperatures had to be used. Another factor in his turning away from platinum was its lack of availability and potential cost. It had been assumed that since the metal had no truly commercial use at that time, little had been done on searching for nature's supplies. Once platinum appeared necessary in his lamp development, Edison sent several men to comb the West for native platinum and platinum ores. Numerous samples were sent to the laboratory for assaying, but no source of commercial interest was found. These discouraging results, plus a warning from his friend Professor Barker of the University of Pennsylvania that it was unlikely platinum could be found in quantity, were also factors in discontinuing the research on platinum filaments.

The high temperature obtained with the carbon arc showed that carbon had a very high melting point. We now know that its melting point is the highest of all elements. Edison's success with highly evacuated glass bulbs indicated there should be no problem with oxidation of the highly flammable carbon and that disintegration resulting from "washing" by convection currents would also be avoided.

Although there was some followup on the work which had been done in 1877–78 on carbonized paper, the effort now was concentrated on the use of a lampblack putty. This composition consisted primarily of lampblack with only enough coal tar—probably 15–20% based on total mix—added to make the mix soft and coherent enough so that it could be pressed into desired shapes. The putty was being used at the time for making flat discs for use in telephone transmitters. Lampblack, as the name implies, is the soot deposited in the chimney of an oil lamp when incomplete combustion occurs. It is nearly pure carbon, and its particle diameter is about 0.1 micron. Lampblack is the ancestor of all carbon blacks. Carbon blacks are made from hydrocarbons by a variety of methods, and they find their major uses as fillers in rubber.

Edison had a small lampblack "factory" on the premises. A battery of kerosene lamps was kept burning 24 hours a day in one of the small wooden sheds. The air inlet of each lamp was restricted so that a highly smoky flame resulted. It was most important that the flame not contact the chimney. One of the night watchman's duties was to harvest the sooty deposits from lamp chimneys each night and to put the lamps again in proper working order.

| After working the lampblack putty in a mortar with a pestle to the desired consistency—Jehl relates carrying out this masticating step for hours—the gunk could be rolled into a slender filament which was then carbonized. As was well known at the time, if vegetable products, tars, low grade coals, and the like, were heated to red-hot temperatures in the absence of air, the volatile ingredients were not only driven off, but the nonvolatile residue decomposed to give gaseous products and substantially pure carbon. Commercial processes then based on carbonization included charcoal from wood and coke from bituminous coal. Besides attempting to use the lampblack in the coal tar mix for preparing filaments, Edison also tried finely ground natural graphite, coke, and charcoal. As binder, he also tried petroleum tars, vegetable resins, and molasses.

Results were encouraging but not surprising. However, now there was no worry about melting the filament. As with platinum, passage of an electric current through the carbon filament while continuing the pumping resulted in gas evolution. The current was intermittently applied while pumping until no more gas was given off at a filament temperature at least as high as that at which the lamp would be operated. In the case of carbon filaments it was not a matter of absorbed gases but rather gases coming from a more complete carbonization owing to the higher temperature of the electrically heated filament as compared with the temperature reached in the carbonization step.

| A major advance was made when a carbonized cotton thread impregnated with lampblack paste was tested. It glowed for over 40 hours and would have lasted longer but Edison kept increasing the voltage. This was October 19–21, 1879, and the 21st is taken as the anniversary for the birth of the electric light *end*

Edison realized it was not the cotton thread as such which accounted for the highly improved life test, but rather a combination of steps all done in an optimum manner. Micro-examination of the carbon structure from cotton thread and Bristol cardboard, both before and after failures as filaments, convinced him that cardboard gave the more rigid and continuous structure. In fact, two identical carbonized cotton-thread lamps were not used for check purposes but to obtain more quantitative data on the conditions of the experiment. A McLeod vacuum gage was sealed in the line between the bulb and the pump, and the carbonized cotton-thread filament was heated until no more gas was evolved at full luminescence. The gage gave a reading in millimeters of mercury equivalent to only one millionth of an atmosphere. This was indeed an excellent vacuum. A bulb containing the other filament was used to measure its resistance in ohms both when cold and hot. The voltage was determined when bringing the light up to 16 candlepower. The amount of energy consumed by the light was determined by a calorimeter by immersing the lighted bulb in a measured quantity of water for a given time and measuring the temperature rise. Although the relationship of $C = E/R$ was known, it had not yet been accepted as a quantitative tool for calculating current once the resistance and the drop in electromotive force through an electrical device were known.

Miniature horseshoe-shaped specimens were punched out from cardboard, but the legs were left attached to a portion of the board. After carbonization the excess carbon was cut off, leaving a shank on each end of the filament to facilitate coupling to the inlet wires. Within a few days after the epic October 21 experiment, a bulb based on a cardboard filament lasted over 100 hours in the life test, and soon 170 hours could be obtained rather consistently. Francis Upton, the mathematician whom Edison had nicknamed "Culture," set up a rack for testing bulbs continuously on a 24-hour day basis. A routine procedure was set up for determining the resistance of each lamp and the voltage required to bring it to 16 candles. Life tests were carried out both on an equivalent voltage basis and on an equal light intensity basis.

President Lowrey of the Edison Electric Light Company was so delighted with the favorable turn of events that he insisted on newspaper publicity and an open-house to demonstrate to the public and Edison's many critics that the light was a success. The famous *New York Herald* article on the new light was published December 21, 1879, and the open-house was held at Menlo Park on the following New Year's Eve. Thousands of people from east and west came on special trains to view the beautiful "globes of fire." Both happenings occurred several months before Edison would have preferred, and it meant that the laboratory had to be turned into a factory the last two months of 1879 to prepare the necessary lamps and auxiliary equipment.

Henry Villard, president of the Oregon Railway and Navigation Company, was a great admirer of Edison. After seeing the light display at Menlo Park in December 1879, he resolved to have the new light installed on the *S.S. Columbia,* a sea-going steamship then under construction for his company. The ship was launched in February 1880 and then brought to New York for loading railway equipment to be shipped to the west coast. While taking on cargo the lighting system was installed with four Edison dynamos and 115 lamps in circuit. The ship sailed early in May and arrived in Portland, Oregon, on July 26, 1880. Chief Engineer Henderson reported that the lights were on 415 hours during the voyage "without one lamp giving out." This was a highly successful field test and no doubt greatly encouraged Edison to move boldly to the lighting of lower New York City.

J. P. Morgan, as a stockholder of the Edison Electric Light Company, installed an electric system for lighting his home in the Murray Hill area of New York City. The steam engine–dynamo setup was installed in his garden, much to the vexation of his neighbors who did not find the engine operation compatible with a good night's sleep. William Vanderbilt, however, was a little more considerate of his neighbors in that he had the generating equipment for the lighting system at his home on Fifth Avenue installed in his basement. Unfortunately, when the lights were turned on for the first time, a short circuit caused a sparking which ignited some tinsel hangings. After the small blaze was extinguished and Mrs. Vanderbilt learned that the terrible noise coming from the basement was a steam engine right in her own home, she ordered the whole system removed.

Although Edison had now chosen the carbon filament, he realized that lights of higher voltage and longer life had to be prepared in order to be competitive with gas lighting. The goal was at least 600 light-hours at 100–110 volts, with filaments of at least 100 ohms resistance. The lights used at the New Year demonstration and on the *S.S. Columbia* were operated at 75–80 volts. They were probably what we would now classify in the 40–50 watt range. The finished light had to be handled very carefully to avoid breaking the filament. The lamps used for lighting the *S.S. Columbia* were packed in cotton, placed in baskets, and taken personally by Upton and his assistant to the ship.

Edison now proceeded to test every carbon-containing material he could get his hands on, including materials which were not cellulosic in nature, such as strips cut from hides, hoofs, and horns of animals, protein fibers, and even hairs from some of his well-bearded employees. Strips of wood from all types of trees were tested, including bamboo taken from a hand fan. The latter proved highly superior to any carbonized fiber or strip previously tested, and a "crash program" was set in motion to test all types available. About 200 kinds of bamboo are known, varying from plants which grow to a height of 120 feet to those of only a few feet. Bamboo is a member of the grass family with "trunks" which are smooth, hollow, and jointed. Branches exist only at the tops. Important as a building material, it is cultivated in parts of the Far East. A certain cultivated variety growing in Japan was found best for making carbon filaments, and a contract was made with a Japanese farmer to supply the desired quantity. A complete switch to bamboo fiber was made in July 1880. A shipment of bamboo filament lights was sent to the West Coast to replace the lights on the *S.S. Columbia.*

We now know that the bamboo fiber was successful not only because it has a high cellulose content but also because it is a continuous fiber between the joints of the stalk. Cellulose is a carbohydrate polymer of empirical formula $C_6H_{10}O_5$. Native cellulose has molecular weights up to about 1,500,000. Although cotton is 98–99% cellulose, the fibers are short, and a continuous thread can be made only by twisting the staple fibers together. The molecules of continuous fibers of cellulose, such as those between the joints of bamboo or regenerated cellulose fibers such as rayon, are intertwined, and perhaps even chemically crosslinked throughout the length of a given filament. On carbonization when the hydrogen and oxygen are driven off, largely in the form of water, the carbon atoms of the massive cellulose molecules tend to remain as intertwined chains. Of course, all polymers do not behave in this manner. Edison carbonized vulcanized rubber with very poor results. Although rubber is made up of giant molecules as is cellulose, when heated the primary result is cleavage of the carbon-carbon linkages rather than dehydrogenation to eliminate the noncarbon portions of the molecules.

Although softwoods contain 40–45% cellulose and hardwoods up to 50%, the other 50–60% consists of materials of low molecular weight such as lignin, hemicelluloses, and a variety of other compounds. In the preparation of paper pulp the non-cellulosic materials are largely removed. Thus the paper and cardboard used by Edison contained largely cellulosic fibers, as does cotton. However, they were relatively short fibers. Carbonized strips of wood proved unsuccessful, no doubt because it is impossible mechanically to separate the pith from the cellulose. In the case of bamboo, a grass, the woody structure is nearly pure cellulose with parallel fibers between the joints. This is particularly true of those fibers near the surface, and it was this portion of the bamboo cane which Edison found best for carbonizing to filaments. In fact, bamboo is the nearest natural substance that is comparable with fiberglass reinforced plastics. Such fiber-reinforced composites are noted for their high strength-to-weight ratios.

Bamboo fiber was used for light filaments for about 10 years, at which time it was superseded by "squirted cellulose." Cellulose had been nitrated as early as 1855 to form nitrocellulose, which when plasticized with camphor became the first man-made plastic. Joseph Swan, the Englishman who had tried and failed to produce a practical incandescent light in the 1870's, learned enough from Edison's success that he formed a company in England and began marketing carbon filament lamps. In 1883 the English Edison Company decided to challenge the Swan Company, but eventually they consolidated under the single name, The Edison and Swan United Electric Company, Limited. Swan pioneered work on the use of regenerated cellulose for lighting filaments. Nitrocellulose could be formed into continuous filaments by dissolving it in a solvent such as ethyl ether and squirting the solution into a non-solvent, such as ethyl alcohol. The fiber had to be denitrated before carbonization by treatment with sodium hydrosulfite, followed by dilute acid and water washing. Edison apparently worked with regenerated cellulose prepared by the Powell process. This process involves dissolving cotton in hot zinc chloride solution, and this is squirted into ethyl alcohol. The fibers were washed free of zinc chloride, wound on drums, and dried. Although this process produced fibers which were suitable for making carbon filaments, it never became important for textile production. In 1889 two new processes were developed. One involved dissolving cellulose in cuprammonium hydroxide to form a soluble copper–cellulose complex. This proved more expensive than the so-called viscose process which consisted of making alkali cellulose using an excess of sodium hydroxide and then forming the soluble xanthate with carbon disulfide as follows:

$$(C_6H_{10}O_5)_nOH(ONa)_2 + 2CS_2 \rightarrow (C_6H_{10}O_5)_nOH(OCSSNa)_2 \text{ (I)}$$

$$\text{(I)} + H_2SO_4 \rightarrow (C_6H_{10}O_5)_nOH(OCSSH)_2\text{(II)} + Na_2SO_4$$

$$\text{(II)} \rightarrow (C_6H_{10}O_5)_n(OH)_3 + 2CS_2$$

Thus, the sodium cellulose xanthate solution is squirted into dilute sulfuric acid, which converts the salt to the free xanthic acid (II). This acid is unstable and decomposes immediately into cellulose. This has become the major commercial process for making regenerated cellulose, now known as rayon, and is also utilized in making cellophane.

The ease of preparing fibers of controlled diameters of any desired length free of non-cellulose material resulted in rayon's replacing bamboo in America and elsewhere in the early 1890's in the manufacture of carbon filament lights. This was actually the first commercial use for the new "viscose silk," as it was then called. The Union Carbide Corporation and other U.S. companies, which currently manufacture high modulus carbon and graphite fibers for aerospace and other uses, have also found rayon to be the best starting raw material in spite of the dozens of new synthetic and natural fibers available. Beginning in 1906, the carbon filaments in incandescent lights were gradually replaced by metal, presently tungsten. Thus, Edison's carbon-type filament was the mainstay of electric lighting for about 26 years. Carbon filament electric lights are still being manufactured by the North American Electric Lamp Company of St. Louis, Missouri, for special uses.

Fastening the fragile carbon filament to the platinum lead-in wires proved to be a vexing problem, and poor connections at these points were the source of many early failures. Initially, a tiny cylinder of hard carbon was used at the junction, with the platinum wire passing into a small hole and the filament end encased in a deep cavity on the opposite side of the cylinder, held in place by carbon paste or fine carbon powder. This method of joining was primarily used with filaments prepared from the lampblack–tar compound.

The method used initially with fibers was to bond short pieces of platinum wire to ends of the raw fiber using "plastic carbon" and to submit the entire assembly to carbonizing tem-

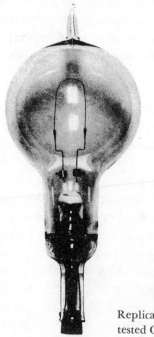

Courtesy Edison
National Historic Site,
West Orange, N. J.

Replica of electric light
tested October 19-21, 1879

peratures. The maximum temperature of this step was well below the melting point of platinum. The platinum wires of the filament assembly were attached to the platinum lead-in wires by "tiny connection pieces." These apparently were small metal cylinders or bars with holes for the wires and screws to hold them firmly therein. This type of coupling was used in the successful experiment with cotton thread, and the arrangement can be seen from the picture of a replica of this light. The electrical resistance of these platinum wires was so much less than that of the carbon filament that the wires warmed up only slightly when the filament was brought to incandescence.

The horseshoe-shaped filaments from cardboard had square shanks on both ends. The finished filament was attached directly to the lead-in wires by small platinum–iridium clamps which "pinched the filament at its shanks, the pressure being regulated by tiny screws." When bamboo was first used, the raw filament specimens were cut from the cane with shanks similar to those from cardboard. Late in 1880 the practice was to plate the shanks of the carbonized bamboo with copper from a copper sulfate solution by electrolysis. Later nickel clamps were used, and finally the clamp was eliminated by welding one end of a copper wire to the platinum lead-in wire; the other end was flattened and folded over the end of the carbon filament. The entire assembly was then copper-plated up to where the filament was joined. This lowered the cost of the light by decreasing the amount of platinum, which then cost $5 per ounce (about $170 per ounce today). By 1886 the filaments were being made without shanks and were bonded directly to the lead-in wires by a graphite paste. This paste consisted of finely ground natural graphite mixed with a little caramelized sugar and gum arabic. The non-graphite portion of the paste was carbonized during the heating of the filament prior to sealing the bulb.

The lead-in wires were coated with glass, and this coating

Courtesy Edison National Historic Site, West Orange, N. J.

Crew of the first factory manufacturing Edison's light. Location is just south of the Pennsylvania Railroad tracks, Menlo Park.

extended a short way within the bulb and served to prevent arcing. The coating also made it easier to seal the wires into the base of the bulb. This technique, plus the fact that very small diameter lead-in wires were used, resulted in excellent bonding between the glass and platinum and made possible maintenance of the near perfect vacuum within the bulb for long periods of time. This was no mean achievement considering that bulbs were subsequently heated and cooled literally hundreds of times when used for lighting.

Edison had expected the Edison Electric Light Company to enter into the manufacture of the incandescent light once its utility had been demonstrated and a demand was established. However, this company had been set up to finance the development of the light and to exploit its patents, and the New York bankers, who now controlled the company, chose to take no further risks. Edison, realizing it was a matter of manufacture or perish, proceeded to set up manufacturing facilities. He financed 80% of the project, and his key employees, Batchelor, Upton, and Edward Johnson financed the remainder.

A two-story clapboard building located about one-half mile from the laboratory across the Pennsylvania Railroad tracks in Menlo Park was used. It had originally housed the manufacturing of the Edison Electric Pen before that operation was taken over by the Western Electric Company of Chicago. To manufacture the electric lamp it was necessary to set up a series of intricate operations from laboratory scale to factory scale. Furthermore, much of the equipment to be used had to be fabricated in the laboratory shop.

It was the first major factory to be powered entirely with electricity. There was not a steam engine on the premises, and electricity was brought in by a line from the dynamos at the laboratory. Electric motors were used to run the blowers used in glass blowing and the carbonizing furnaces, to rotate the glass-annealing machines, operate a saw to cut the bamboo, power the Archimedes screws used in lifting the mercury to feed the vacuum pumps, and to light and test the lamps.

The bamboo was sawed into required lengths, split into splints, planed down to proper thickness, and then cut by a special apparatus into raw filaments. It was steamed and dried prior to carbonization. Nickel molds of approximately 3¾ x 1¼ x 3/16 inch outside dimensions were originally used for this operation. They were hollowed out in an oblong fashion, and only one filament was placed in each mold. The molds were placed one upon another in a large graphite container packed with carbon black. Heating was carried out in a fire-brick furnace using coal gas produced on the premises.

In July 1881 John W. Howell, who had just graduated from Stevens Institute of Technology, was hired to work at the lamp factory. Howell worked in the field of lamp production and utilization his entire life and later became an outstanding authority on the subject and co-author of an excellent book on the history of the incandescent lamp (*10*).

When Howell first came to Menlo Park the bamboo filaments were being heated in individual carbon boxes instead of nickel. As indicated in a letter to Jehl when the latter was writing his "Menlo Park Reminiscences," Howell wrote as follows relative to improvements made in the carbonization step in 1881 (*11*):

We soon learned to pack them in peat moss which was first roasted. This shrunk the same amount as the bamboo and kept the fibers in shape. When we left off the shanks, everything was simpler and we packed large numbers in peat and carbonized them. We carbonized them in two operations, the first was done in iron boxes—up to 600° Fahrenheit. These were heated very slowly, for if we went too fast the tar which was distilled out of the bamboo and peat would come too fast and stick all the carbons together. The first heat took from six to eight hours. The second carbonizing was done in graphite boxes and they were run up as fast as possible to a white heat. The first was called the preliminary and the second the final heat. All the shrinkage took place in the first heat, the mass of peat and fibers shrunk away from the box leaving an empty space all around. This mass of peat and fibers was put bodily in the graphite crucible or box. Many thousands could be put in one box.

It is likely that the "white heat" mentioned by Howell was more nearly a cherry red and not over about 1200°C. Since the electric furnace had not yet been invented, Edison was restricted to maximum temperatures available from a practical gas-fired furnace. This temperature limitation made the electrical heating of the filament in the light bulb prior to sealing a matter of completing the carbonization step.

The glass bulbs were originally hand blown at the factory from 1-inch tubing, but soon they were purchased from the Corning Glass Works where they were hand blown with glass directly from the furnace. Later they were blown in molds which resulted in better uniformity. The glass stem containing the lead-in wires sealed into the neck of the bulb was handmade until 1901 although Edison had developed his "Dufunny" machine to seal the stem to the bulb in 1889.

The time-consuming and critical operation was evacuating the lamp, expelling gases from the partially carbonized filament and the moisture adsorbed on the inner glass surface, and sealing the bulb at the tip while maintaining the high vacuum. Two hundred simplified Sprengel pumps which had been constructed by glassblowers at the laboratory were installed in rows. They were arranged in units with an iron reservoir to collect the mercury at the bottom of the pumps; the mercury was raised in an Archimedes screw pump to an elevated tank which fed it to the individual vacuum pumps. Each pump had a resistance box to control the current used to expel occluded gases from the bulb. A circular gas flame was used for the final heating of the exterior of the bulb to remove adsorbed gases and moisture. The bulb was sealed in place by hand. Phosphorus pentoxide was used to absorb moisture from the gases in order to keep the mercury and pump dry, and gold leaf served to absorb mercury vapors. Edison was aware of the toxic nature of mercury and appears to have done a good job in protecting laboratory and factory personnel from health hazards while handling large quantities of this liquid metal.

Anyone who has carried a new process from the laboratory to pilot plant, and then to plant operation, well knows the headaches involved. Starting this lamp factory was certainly a tremendous undertaking. Think of the potential breakage and process problems related to setting up 200 glass vacuum pumps to be operated by unskilled personnel circulating large quantities of mercury and evacuating and melt-sealing glass bulbs to 0.000001th of an atmosphere. The reliable Batchelor set up the lamp factory and managed its operation until January 1881 when Francis Upton took charge, "Mr. Batchelor's health having given out." The lamp manufacturing project, however, was an unqualified success. Approximately 50,000 lamps were produced the first year. Edison had to agree to a fixed price per lamp with the Edison Electric Company and this was set at 40¢. The first year the cost was $1.10 each; the second, 70¢; the third, 50¢; and the fourth, 37¢. The manufacturing operation was moved to larger quarters in Harrison, New Jersey in April 1882. The volume had increased so drastically that the 37¢ cost in the fourth year made possible the recovery of losses of the first three years. Production costs were about 30¢ per lamp in 1885 when 139,000 lamps were made. In the late 1880's the company's annual output approached a million lamps, with costs per unit down to 22¢.

Edison's Patent 223,898 entitled "Electric Lamp" was applied for on November 4, 1879 and was granted January 27, 1880. It covers less than two pages and has only four claims. One of the most important patents of all time, it is not only a model of brevity, but also one of completeness of disclosure. Claims 1 and 2 describe the invention:

1. An electric lamp for giving light by incandescence, consisting of a filament of carbon of high resistance, made as described, and secured to metallic wires, as set forth.

2. The combination of carbon filaments with a receiver made entirely of glass and conductors passing through the glass, and from which receiver the air is exhausted, for the purposes set forth.

Although substantial changes have been made in the electric lamp as invented by Edison, much of his original concept and design remain. Incandescence has been universally accepted as the best form of lighting. Even after the light had been demonstrated, most experts in the field thought it could never compete economically with combustion as a source of light. Imagine our greater air pollution problems today if city streets were illuminated by arc lights and homes and offices by burning gases and oils. The source of light in electric lamps is still primarily a high resistance filament. Edward Acheson, a former employee of Edison, developed an improved resistance-type electric furnace in 1895, which produced temperatures approaching 3,000°C. Beginning at about 2,200°C amorphous carbon rearranges to the crystalline molecular structure known as graphite. If Edison had had the electric furnace available to carbonize filaments, he could have made a light which could have been used at a higher temperature and therefore with greater efficiency. The filament of the Edison lamp had a temperature of about 1,775°C. Higher temperatures could not be used since under the high vacuum conditions sublimation of the carbon became appreciable.

The Gem light bulb developed by the General Electric Company in 1905 had a graphite filament. A tantalum filament lamp was developed in 1906 and the tungsten type shortly thereafter. The tungsten filament was first pressed out from metal powder, but by 1911 a process had been developed for drawing tungsten wire. With the advent of the tungsten filament, it was found advantageous to change from a vacuum to a gas-filled lamp, usually a mixture of argon and nitrogen, in order to allow a higher temperature to be used without sublimation of the tungsten. Present day incandescent lights have tungsten filaments supported by molybdenum wires and operate at filament temperatures in the range of 2,400°–2,600°C.

The lamp is still an all-glass enclosure. As platinum became increasingly more expensive, major efforts were made by the industry to find a substitute material for the lead-in wires. In 1911 a composite wire consisting of a silver-plated nickel–iron alloy core with a platinum sheath was found applicable. Shortly thereafter Dumet wire, a nickel–iron core with a copper sheath, found universal use for this application. The screw-cap type connection for fitting the bulb into the socket remains as it was in 1880, as do the electrical connections to the outer circuit *via* the rim of the bulb and a center contact insulated from the rim. Few of our intricate articles of commerce have changed so little from their initial design.

The development of so-called baked carbon is credited to Carré, a Frenchman, who in 1876 carbonized a composition containing coke, lampblack, and sugar syrup. Whether Edison knew of this development when he carbonized lampblack-coal tar filaments two years later is not known. Edison was certainly the first to make carbon fibers; he discovered that bamboo was the best natural fiber and rayon the best synthetic to employ as precursor materials, and they still are today. Although certain other synthetic fibers such as polyacrylonitrile are now being used to some degree to manufacture carbon and graphite fibers, the primary production in the United States in 1970 is still from rayon.

Probably no scientist or engineer has ever been the subject of ridicule as was Edison during the time the electric light was under development. The following article appeared in the (London) *Work and Design* magazine in 1880 and was reprinted without comment in *The Operator,* the foremost electrical publication in the United States at that time (*12*):

Edison's Electric Railway

Mr. Edison is to the front again with another idea. This time he does not offer us a new anaesthetic, a remedy for baldness, or a patent cradle rocker. He has "invented" an electric railway; but it is somewhat curious that his invention follows, with a suspicious interval, the previous announcement, in these columns and elsewhere, of a similar invention by Dr. Werner Siemens, of Berlin. Edison's invention in itself, is neither new nor good, but we do not doubt that he bears in mind a certain adage, which we do not care to repeat here; and hence the publicity and excitement following closely on the heels of the great gold-creating scheme, from which our Transatlantic cousins are suffering. Mr. Edison's latest bantling, his railway scheme, has been tried at Menlo Park on a length of half a mile of rails laid without regard to level or curves. The motor is one of "Edison's generators," and the electrical current is supplied by a local engine, the rails being insulated and used as conductors. The results of the trial are recorded with more adjectives than nouns, and a kind of vague precision that is wholly charming. The locomotive "pushes ahead with wonderful energy, and at a break-neck speed, up and down the grades, over humps and bumps, at the rate of twenty-five miles per hour." We quote from a professedly scientific report, which adds that the great successor of Barnum realizes 70 per cent of the original energy, and apparently hopes to double it—we shall not be astonished if he does.

Further particulars are promised, but we do not expect that the inconstant "inventor" will achieve anything on his present track, for it is safe to prophecy that he will very shortly produce a diaphote and claim to be the true and sole inventor, and then the latest scheme will be abandoned—as many of his schemes have been—as good-for-nothing.

Such hysterical criticism was not restricted to trade magazines; similar comments came from many who were prominent in scientific and educational circles. When Edison announced in 1879 that he had developed a dynamo capable of converting 90% of mechanical energy to useful electrical power, this was branded the greatest fraud since perpetual motion. Based on the energy output of primary batteries, it was accepted by most that the maximum efficiency of an electrical generator was 50%. Even Dumas, the great French chemist, felt that fragile carbon could not stand an incandescent temperature long without disintegrating. Subdivision of the electric current without change of pressure (voltage) as the distance from the source increased, was deemed impossible. From the start, Dr. Henry Morton, President of Stevens Institute and one of the noted scientists of the area, forecast that Edison's efforts would come to naught. Edison was pictured as an uneducated former newsboy who knew no better than to ignore established scientific principles. His premature announcement in the fall of 1878 of his successful approach to the problems of subdivision and incandescent lighting, and the fact that it was over a year later before he could demonstrate it, no doubt fed fuel to the criticism.

Edison, in his early thirties at the time, proved to be a true scientist when he accepted only what he could demonstrate by experiment and ignored what older, and supposedly more learned, people said if they did not agree with his experimental data. He did not become bitter and then gloat over his former critics once the electric light lived up to his predictions.

Chapter 3 Accessories for the Electric Light

ON THE AFTERNOON of September 4, 1882, a group of men including Thomas Edison assembled in the offices of the famous banker, J. P. Morgan, at 23 Wall Street, New York City. The occasion marked not only the inauguration of electric lighting from the first central station to numerous customers but also the first commercial installation of any type, including telegraph and telephone, where underground wiring was used. Edison, in his usual all or nothing manner, had chosen not only the largest and foremost city in the country to demonstrate his new "hot hairpins in bottles," as his new electric lights were often called, but the most important part of that city. The one square mile of his franchise included the financial area now known as Wall Street, and the editorial offices of the *New York Times* and *New York Herald*. It was obvious that Edison was going to impress the proper people with his new lighting system, with no thought that in case of failure it would have been much better to have had the first installation in a remote area in that or some other city.

As three o'clock approached—the time scheduled for throwing the switch at the Pearl Street Station to send electric current through the underground mains—Edison drew out his watch which had been synchronized with that of John Lieb, chief electrician at the station. "A hundred dollars they don't go on," volunteered one of the members of the board of the Edison Electric Light Company. "Taken," said Edison. Moments later, the lights in the office came to a steady, mellow glow.

It was indeed no mean achievement to have gone from lighting the Menlo Park laboratory in January 1880, to lighting the most important financial and business area of New York City two and one-half years later. Such a time interval is typical, for example, in carrying a new chemical process to commercial scale operation. However, in Edison's case it was more than just a matter of carrying his development from the laboratory to commercial scale; in the interim he had to make dozens of new devices and other equipment, none of which was available on the market and, in fact, some which had not yet been developed. Additional problems included financing, a franchise to be obtained from city officials, and, most important, the capability to install a new lighting system which would be competitive with the gas lighting already in use in the area.

Next to developing the light itself and improved dynamos for generating the current, both of which Edison felt were well on the road to success in early 1880, the major problem was to deliver electrical energy of uniform characteristics from a central point to an area up to several square miles. Edison, as did the pioneers of arc lighting, assumed from the start that metallic copper would be used to transport the electric current, as well as to wind the armatures and electromagnets of dynamos and motors. Copper was used commercially as a conductor of electricity as early as 1811 in Germany (1). As shown in Table I, copper is the best conductor, after silver, of all metals at ambient temperatures when compared on a cross-sectional basis.

Silver and gold were too scarce and expensive to be considered, and aluminum was not commercially available in 1880. However, copper supplies were plentiful, and it had been used since Neolithic man discovered and first used it during the late Stone Age, probably about 8000 B.C. By 6000 B.C. the

TABLE I.   CONDUCTIVITY OF METALS

	Volume Resistivity, 0°C
Silver	1.47×10^{-6}
Copper	1.56×10^{-6}
Gold	2.20×10^{-6}
Aluminum	2.66×10^{-6}
Iron	9.06×10^{-6}
Platinum	$10.9 \ \times 10^{-6}$
Lead	$20.4 \ \times 10^{-6}$

ancients had learned how to melt and cast it, and by 3000 B.C. the Egyptians and others could reduce carbonate and oxide ores of copper by heating with charcoal. Since copper so prepared was contaminated with various other metals, alloys appeared and marked the beginning of the Bronze Age when tools and weapons were made from the copper–tin alloy. In spite of low cost aluminum today, as well as literally hundreds of alloys of various metals, copper continues to be the "workhorse" of the electrical industry, as evidenced by the fact that of the more than three million tons used annually, over 50% is consumed in various electrical applications.

The copper-wire industry was well established by 1880, having grown phenomenally as a result of the telegraph industry. Although steel and galvanized iron wires were also used, primarily because of their greater tensile strengths, the bulk of the telegraph lines were bare copper supported on poles. Edison was able to purchase bare copper wire and rods of desired dimensions. Although by 1881 copper wire became available specifically for electric light purposes, it was of varied quality. Edison, however, had facilities to test conductivity of copper wire samples submitted by manufacturers.

Anyone who has visited the Edison National Historic Site in West Orange, New Jersey has most likely observed a block of copper, $1 \times 1 \times 1$ foot in dimension on a wooden pedestal in the library. This was presented to Edison by the copper industry in recognition of what his pioneering developments had meant to the growth of that industry. When Edison was informed that representatives of the copper industry wished to give him a suitable award, he suggested a cubical block of copper. Edison, always the practical joker, commented, "You fellows are always talking about a cubic foot of copper weighing 551.8 pounds. I would like such a cubical sample." He knew well that it was necessary to work copper in order to remove gas bubbles. Thus, it was necessary to mold 11 cubical castings before one with a porous-free surface was obtained. It was presented to Edison even though it weighed only 486 lbs, less than 90% of the theoretical weight based on the specific gravity of copper.

The question raised by Edison's critics was not whether copper would be a suitable conduit for electricity from source to customer, but rather would there be enough copper available to supply even the large cities if electric lighting were generally accepted. Arc lights were wired in series, with each dynamo acting independently and providing current for one or more lights. When Edison announced that his light was to be of high resistance, individually controlled from a source of constant pressure, it was visualized that each light would be wired directly to the primary source. Such conductors might either be separate or a tapering conductor, massive in cross-section near the source, and decreasing in diameter after each individual light or group of lights. Such a setup was known as the tree system and was used in the first isolated lighting installations. In mid-1880 Edison developed his so-called feeder system whereby the main distribution lines would be supplied at central points by small diameter feed lines in which a predetermined drop in voltage, say 120 to 110 volts, would be taken (2). This reduced the copper requirement for total transportation lines in a one-half square mile area to one-eighth of that when using the tree system and even greater savings in larger areas served by a single central station. Later this was supplemented by the three-wire distribution system which, because of higher voltages in the main lines, made a further 62½% reduction in copper requirements as compared with the two-wire system (*3*). This reduction in copper requirements to an acceptable economic level by the use of additional conduits of lower diameter would not have been feasible unless highly effective electrical insulation for the conduits was developed. Such insulation was not available in 1879 when Edison's famous carbonized cotton-thread lamp ushered in the age of incandescent lighting.

Metals in general are good conductors of electricity, with the current moving as free electrons. It was recognized early that a metal which carried electricity should contact only insulating materials, whether gases, liquids, or solids. Since air was known to be an insulator, telegraph, telephone, and early electric wires were simply held in air, attached to glass or porcelain supports mounted on poles, roof-tops, trees, or the like. Although liquids, such as mineral oils, are widely used today to insulate cables, transformers, circuit breakers, and the like, Edison never made use of them. However, solids are the most effective insulating materials, and these generally were the materials Edison investigated for insulating electric conduits.

Just as conductors such as copper have some resistance to the flow of electricity, all insulating materials conduct some current. However, considering that a good insulation material has a volume resistivity of 10^{15}–10^{18} ohm-centimeters, its conductivity compared with that of copper differs by an order of 10^{22}. Insulators conduct current by means of ions which carry negative or positive charges, and since organic compounds have primarily covalent bonds between atoms, this class of chemical compounds represents the best insulators. Moisture and inorganic salts, acids, and bases may dissolve in an organic material, thereby causing it to be a poor insulator. Thus, a solid organic material which absorbs little or no moisture makes the best insulator. Hydrocarbons, particularly those of high molecular weight, should be the best insulators. Therefore, natural rubber and gutta percha were two materials used at an early date to insulate telegraph cables (4).

Although there had been various electrical wire insulations used on an experimental and developmental basis, the only insulated wire on the market in 1880 was cotton-covered wire for the telegraph and for constructing electromagnets. Wood was considered a good insulator in those days, and in the case of the first commercial installation of the Edison lighting system —i.e., that of the S.S. *Columbia* in the spring of 1880—bare wires were simply tacked to wooden supports. However, Edison realized that improved insulations would have to be developed, particularly since he had decided to go underground in his first central lighting station in New York City.

The open-house at the Menlo Park laboratories the end of December 1879 was held primarily to display the new light. In some respects it had the aspects of a medicine show. A total of about 100 lights were mounted in all types of positions to show that they did not have to be upright as do gas flames and candles. One lighted bulb was even immersed in water. An assistant stood by a switch and turned a light on and off several times a minute. A sewing machine was run by a small electric motor using current from the same line as that for lighting.

Edison correctly realized that before lighting city homes, offices, factories, and streets with power from a central station, a full-fledged pilot-plant operation would have to be carried out. As in the case of nearly all pilot plants, the objective of the installation was not only to test the technical feasibility of all aspects of the future commercial installation but also to get reliable cost data. The Menlo Park area was chosen for the full-scale test. Approximately 425 lights using bamboo filaments were installed in the laboratory buildings, the Jordan boardinghouse across the street, the stairs of the railroad station, and rows of lights suspended from poles in the fields to create imaginary streets. The total area in this test was approximately one-half square mile, utilizing six miles of main lines. Ten dynamos, each with a capacity of 60 lamps, were built to supply the power, plus an extra one to provide current to the magnets. Arrangements were made with Western Union to supply a gang of men to install the wiring and poles for the street lighting.

A procedure was set up to test current losses in short lengths of wire laid underground. A Daniell cell was used as current source, and a telegraph key and sounder were used to test whether or not current flowed; if it did, loss of current was followed by a Wheatstone bridge. To keep expenses low and to obtain a more severe test of the insulation, the underground wires were laid in wooden troughs or in no enclosures at all. A control test with bare copper wires in grooves in a wooden trough resulted in severe current leakage under wet conditions. Studies were made by pouring various insulation gunks around the wires held in grooves in the boxes. Coal tar was found to be poor, and this was traced to the high acidity of the tar.

Wilson S. Howell, one of Edison's young assistants at the time, recalled in the 1930's how the underground insulation problem was solved (5).

How to insulate these wires was a knotty problem. Mr. Edison sent me to his library and instructed me to read up on the subject of insulation, offering the services of Dr. Moses to translate any German or French authorities which I wished to consult. After two weeks' search I came out of the library with a list of materials which we might try. I was given carte blanche to order these from McKesson & Robbins and within ten days I had Dr. Moses' laboratory entirely taken up with small kettles in which I boiled a variety of insulating compounds. The smoke and stench drove Dr. Moses out.

The results of this stew were used to impregnate cloth strips, which were wound spirally upon No. 10 B. W. G. wires one hundred feet in length. Each experimental cable was coiled into a barrel of salt water and tested continually for leaks. Of course, there were many failures, the partial successes pointing the direction for better trials. These experiments resulted in our adopting refined Trinidad asphaltum boiled in oxidized linseed oil with paraffin and a little beeswax as the insulating compound to cover the bare wire cables, which had been previously laid alongside trenches throughout the streets of this little Jersey village. Barrels of linseed oil, bales of cheap muslin and several tons of the asphaltum were hauled in, two 50-gallon iron kettles were mounted on bricks, and the mixing operation was soon progressing in a big way. Through the pot in which this compound was boiled, we ran strips of muslin about 2½ inches wide. These strips were wound into balls and wrapped upon the cables. After the man who served these tapes upon the cables had progressed about six feet, he was followed by another man serving another tape in the opposite direction, and he in turn by a third man serving a third tape upon the cable in the direction of the first winding. After the cables were all covered with this compound and buried, the resistance to the earth was found to be sufficiently high for our purpose.

Of the four ingredients used in the proprietary insulation compound described by Howell, the paraffin wax has the best insulation characteristics. Those of asphalt are also reasonably good when purified of inorganic salts. It has the very favorable characteristic of being a hard solid at underground temperatures but becomes quite fluid at 200°F or so, which would permit good handling properties. Because of its aromatic nature, in contrast to the highly linear molecular configuration of paraffin wax, asphalt changes viscosity rapidly with changes in temperature. Also, as a result of its aromatic chemical composition, it would wet out the copper wires well and thus tend to eliminate adhered moisture and air. Asphalt contains sulfur compounds, some of which are labile enough to react with copper at hot application temperatures. This would result in strong bonding of the insulation to the wires. As is well known, asphalt contains compounds which make it immune to attack by soil bacteria. Oxidized linseed oil hardens much like a film of linseed oil exposed to air. Thus, this liquid ingredient would make the mix a pliable compound, but unlike a non-reactive liquid, such as a mineral oil, it would harden in time. The small amount of beeswax probably acted as peptizing agent— that is, it made the overall blend compatible. From the standpoint of insulation and physical characteristics, the four-component mixture would appear to have had merit for the desired use. A suitable electrical insulation must have the desired physical and handling properties and must be a non-conductor. Cost was also a very significant factor, considering the large quantities of insulation used when the wires were embedded in wooden troughs or metal tubes.

Howell's compound proved to be highly effective. Although most of the underground wiring at Menlo Park was abandoned after the pilot test, a section was recovered in 1933 and found still to be a well insulated conduit. It had been buried 12 inches underground with only the muslin-impregnated wraps. Underground wiring at Edison's winter home in Fort Myers was installed in the spring of 1887, and 10% of the original underground wiring remains today, more than 80 years later. The rest was replaced in 1968. The insulation compound was used in metal tubes containing copper rod conduits well into the 20th century. It was also used to prepare insulation tape for insulating wire splices and the like.

The extensive lighting system at Menlo Park was operated for about three months. The well lighted village attracted considerable attention from passengers travelling through on Pennsylvania Railroad trains. One reporter commented that nothing was more beautiful than "Edison's fairyland of lights" on a snowy night. Considerable experience was obtained with a number of dynamos which furnished current to a common system. Based on a series of 12-hour runs, it was found that 0.4 lb of coal would furnish enough energy *via* the steam engine–dynamo conversion to light an incandescent bulb for one hour. The bulbs were 16 candle power and today would be rated about 95 watts.

The cost data plus the successful operation of the extensive lighting installation at Menlo Park convinced the directors of Edison Electric Light Company that they were ready to move to commercial central station lighting. The New York City mayor and aldermen were invited to Menlo Park on December 20, 1880, and the pilot operation demonstrated and explained to them. The technical briefing was followed by a banquet in the upper room of the laboratory where luxurious foods and wines were catered by the renowned Delmonico Restaurant. It was said that the food and drink might have impressed the political guests more than the technical display. The gas companies operating in New York strongly opposed Edison's proposal to set up a central lighting station in that city, which is not surprising since he had predicted that his new light would displace gas lighting. However, J. P. Morgan, who held a considerable financial interest in Edison Electric, apparently was able to apply the proper kind of "persuasion" to the right people, and on April 19, 1881, an operating franchise was granted covering one square mile of lower Manhattan.

Edison did not wait for the official franchise to be granted before getting ready to manufacture suitable underground wiring. As in the case of the light bulbs, he received no financial help from the Edison Electric Light Company for manufacturing underground conduits. In January 1881 he formed the Edison Electric Tube Company, and operations began at 65 Washington Street, New York, on February 14. The Pearl Street Station was wired with the feeder-main system, but the mains constituted two wires. The conductors, half-round copper rods 20 feet 6 inches long, were originally separated by cardboard spacers. This arrangement was soon changed to one whereby each rod was wrapped with a Manila rope, after which the two were wrapped together with flat sides toward each other, using a third rope. The two rods were then inserted in a 20-foot cast iron pipe. The hot asphalt–linseed oil–paraffin wax–beeswax compound was pumped through the pipe, filling all empty space. It solidified on cooling and served to insulate the conductors from each other as well as the enclosing pipe.

Insulation for indoor wiring went through a series of developments. Wood was first thought to be an adequate insulator. However, in wiring homes, the metal gas fixtures were often used to support the light sockets, and good insulation was required. Cotton-covered wires and cotton impregnated with shellac, paraffin wax, rosin, and the like were used. The asphalt compound was hardly suitable where contact might be made with wallpaper or even curtains. Initially no attempt was made to conceal the wires. In 1881 insulated indoor wiring encased in cardboard tubes was developed. This was the forerunner of the familiar BX and Romex cables.

One may wonder why Edison did not give more attention to rubber for insulation since it was one of the early materials studied for insulating underwater telegraph lines and today is used widely to insulate electrical wiring. The only rubber available in 1880 was low grade natural rubber obtained from trees which grow wild in Brazil. It was expensive, about 50¢ per lb, and processes developed at that time for coating wires were expensive and unsatisfactory overall. Rubber was usually applied by dipping the wires into a solution of rubber followed by evaporation of the solvent, or it was applied in the form of rubberized cloth. Although the technique of extruding rubber around wire by means of a press utilizing a heavy steel cylinder and steel ram had been developed, it was a crude operation compared with present-day continuous methods which operate in conjunction with continuous vulcanization of the rubber-coated wire. The vulcanization process at that time was extremely slow, requiring hours. Although in his original patents, Goodyear disclosed basic lead carbonate as an accelerator of sulfur vulcanization, only inorganic materials such as lime, magnesia, and antimony sulfide were used for this purpose in the 19th century. In 1906 Oenslager of the B. F. Goodrich Company found that condensation products of aldehydes and organic amines, such as hexamethylenetetramine and substituted guanidines, greatly accelerated the sulfur vulcanization of rubber. This epic discovery made possible the vulcanization of rubber in a few minutes or less in the case of thin layers of rubber-coated wire where the temperature of the rubber can be raised rapidly to vulcanization temperature.

The most important reason why Edison did not concentrate on rubber for electrical insulation was no doubt the fact that it deteriorated within a few months when exposed to air, particularly when exposed outside. This deterioration is caused by chemical attack of the double bonds between carbon atoms of rubber by oxygen, and even much more so, by ozone. Ozone, a highly reactive gas containing three atoms of oxygen per molecule in contrast to two for atmospheric oxygen, is produced photochemically in the atmosphere by the action of ultraviolet light on oxygen. Although the atmosphere at ground level contains less than one part per million of ozone, it is present in considerable concentration at high altitudes. However, it is primarily this highly active chemical which causes rubber exposed to outdoor atmospheric conditions to crack and become brittle. Since ozone tends to be adsorbed on solid surfaces where it reacts chemically with them or spontaneously decomposes to oxygen, there is much less ozone inside buildings than outside. It is formed in relatively high concentrations when air is subjected to an electric discharge and often is formed during electrolysis. Thus, insulations of electrical wiring and devices are likely to be subjected to much higher concentrations of ozone than the adjacent area. The so-called "electric smell" around electric generators and motors, or even outdoors during a severe electric storm, is the odor of ozone.

Carbon-black filler in rubber is an excellent protective agent against outside aging—that is, against ozone cracking and oxidation by oxygen catalyzed by ultraviolet light. Rubber containing carbon black is apparently protected by screening of the light rays and the selective attack of ozone on the finely divided carbon. Its protective effect on rubber was not known in the 1880's since carbon blacks were not widely used in rubber compounds until about 1912. Organic additives, which function as antioxidants and antiozonants for rubber, even in concentrations on the order of one percent, are of rather recent vintage.

Edison realized early that electric wiring had to be protected from possible overload. Even at the open-house festivities held at Menlo Park the end of 1879, fuses—or safety wires as they were first called—protected all electrical equipment. This proved to be fortunate since a visitor from a Baltimore gas company attempted to short out the whole system by means of an insulated wire which he had run through the sleeves of his coat with his hands concealing the ends. When he slyly put this jumper in contact with two circuit wires, only four lights went out and they were protected by a fuse. The culprit was roughed up a bit and escorted from the premises.

As in the case of the light, a screw shell form of socket was used for the safety fuse. The practicality of this choice is evident from its use up to present day. The shells for the fuses were first made of plaster of paris, then of wood, glass, and finally, porcelain. The safety wire was first made with copper, then lead, and later with an alloy of 60% lead and 40% tin. In March 1880 Edison filed his patent application entitled "Safety Conductor for Electric Lights" which was granted the following May as U.S. Patent 227,226. In it he stated:

This safety device consists of a piece of very small conductor interposed in the main conductors of a house or in the derived circuit of a lamp. Preferably, one is interposed in the circuit of each lamp or other translating device. This small conductor has such a degree of conductivity as to allow readily the passage of the amount of current designed for its particular branch, but no more. If, from any cause whatever, an abnormal amount of current large enough to injure the translation devices or to cause a waste of energy is diverted through a branch, the small safety wire becomes heated and melts away, breaking the overloaded branch circuit.

The Edison safety plug became a standard all over the world. The fire underwriters required it as a part of all installations of electric current. Originally this fuse was designed only for 110-volt current, but shortly afterwards one for 220-volt circuits was developed. At one time the fuses sold as low as 5¢ each and were available in almost any kind of store.

A basic problem in subdividing the current from a central power plant to numerous customers was how each customer was to be charged. The only previous electric lighting systems were the arc lighting of streets and Edison's isolated lighting plants. In case of arc lighting, the city or customer was charged so much per light, and the electricity was generated only during hours of darkness. For an isolated plant, such as the *S.S. Columbia,* the cost was a contractual arrangement with the one customer. It was generally assumed that customers who used electricity from a central station would pay so much per light. However, at the time his research began in late 1878 on replacing gas lighting with electric lighting, Edison visualized that the

84

customer would have to be charged on the basis of the electrical energy used, whether it was for lighting or for power.

The fact that Edison took out 22 United States patents on meters for measuring electricity is ample evidence that he spent considerable time and money developing suitable means for measuring the amount of electricity used by individual customers. The meter successfully developed and used up to about 1896 was based on electrolysis and became known as Edison's chemical meter. It seems paradoxical that a force like electricity, which has no weight, was originally marketed by weight. Obviously a meter based on electrolysis was applicable only for direct currents, and those of a known and uniform voltage.

Edison was an ardent student of Faraday, the famous English physicist-chemist who did so much to establish the relationship between chemistry and electricity. Faraday discovered that when an electric current is passed through an electrolyte, the chemical decomposition action is constant for a given quantity of electricity. He also determined that the electrochemical equivalent of an element is numerically the same as the chemical equivalent. Thus, the same quantity of electricity liberates 107.88 grams of silver, 23 grams of sodium, or 1.008 grams of hydrogen since these values are their respective atomic weights and each element has a valence of one. For a divalent metal such as zinc, twice the amount of electricity is required to liberate one atomic weight—*i.e.*, 65.37 grams. The merit of using electrolysis to measure current passed is that the amount of electro-decomposition is not affected by the voltage, the time interval the current passes, the temperature, or the concentration of the solution.

However, Edison knew that he would have to avoid the chemical meter which could act as a battery and redeposit the metal from the cathode when current was not being passed. The principal problem was to avoid having the meter act as a storage battery in reverse to the electrodeposition from the outside current. Also, there was the problem of the resistance of an electrolytic cell decreasing with an increase in temperature. Since the meter was to operate from a shunt which carried the main current and since the percent of total current passing through the meter had to remain constant so that the total ampere-hours used could be calculated, the meter had to be of constant resistance. This problem was solved ingeniously. Since the resistance of metals decreases with temperature rise—*i.e.*, the opposite of aqueous solutions—Edison put a compensating resistance of copper wire in the circuit with the electrolytic cell. Copper wire was used since its resistance varies greatly with change of temperature and quantitative data were available on this change. Thus, once the cell was standardized and its change of resistance with change of temperature determined, the length of copper wire of a given diameter for compensation purposes could be calculated. Since only about 0.1% of the current used by the customer went through the meter, the loss of current resulting from heat generated was insignificant.

As to the problem of reversed deposition, it was known that silver shows no tendency in this direction and was in fact used at that time as a standard cell to measure current density. Copper electrodes perform poorly in this respect, and lead would be very poor since it is the best metal for use in storage batteries. After considerable experimentation high purity zinc was found to be the most practical metal to use in the chemical meter from the standpoint of chemical behavior and cost.

The original Edison meter, which became commercially available in early 1882, utilized two zinc plates, 3×1 inch and $\frac{1}{4}$ inch thick. They were held 0.24 inch apart by two ebonite (hard rubber) blocks, one at each end of the plates, held in place by an ebonite screw coupling the two plates at the center. The pair of electrodes was placed in a glass jar containing zinc sulfate. Initially there were two such cells in the meter box, one wired in such a way that it received only one-third as much current as the other. The idea was that the one would be used for monthly readings and the other used quarterly for check purposes. The meters proved to be so accurate that the quarterly meter was discontinued. However, the meters continued to have space for two cells, but of equal resistance, so that they could be checked if desired. The shunt wire within the meter was of German silver and was of large cross section to be of low resistance. German silver, an alloy of copper, zinc, and nickel, was chosen because its resistance varies little with temperature. Since over 99% of the current went through the shunt, any change in the resistance of this bypass wire arising from temperature change was negligible when calculating current used.

An electric light bulb was included in the base of each meter cabinet wired in circuit with a thermostat set to turn on the light at freezing temperatures. Although zinc sulfate is highly soluble in water and the concentrations used were high enough to prevent the water from freezing even at subzero temperatures, the heat from the light served to avoid highly viscous solutions at subfreezing temperatures.

The chemical meter was serviced monthly; a cell was removed, taken to the central station, the plates were removed, washed, and dried, and the amount of zinc transferred from anode to cathode was determined by weighing each on a precision balance. As a whole, the meter was cheap to manufacture, easy to install, and was accurate. In 1888 the Boston Edison Company reported they had 800 chemical meters in use. They were serviced by two men and three boys at an annual cost of $2,500.

The Edison meter was the subject of technical lectures presented both in the United States and abroad. By today's standards the meter would appear complicated and expensive to maintain. However, it was one of the pioneering developments in electricity in that for the first time a technique was developed for measuring electricity sold.

Edison also invented a motor type meter, U.S. Patent 242,901, which operated on the principle of recording the number of revolutions of a miniature motor. Increase in current would cause the motor to run faster, and vice-versa. Edison did not reduce this meter to commercial practice. The chemical meter could not be used with alternating current. As this type of electricity entered the American market in the late 1880's, various mechanical meters were developed by others. Elihu Thomson of the Thomson-Houston Company made major contributions perfecting such meters. The New York Edison Company used the chemical meter exclusively until October 1896 at which time they had 8,109 in use. After that they purchased only mechanical meters, but even at the turn of the century some of the chemical type were still in service.

Edison's development of dynamos of improved efficiency and size was a critical factor in his being able to market electricity at a price competitive with gas. He was the first to go to direct drive—that is, a common shaft for the steam engine and the dynamo armature. Also, he was the first to use mica for insulating the bars of the commutator. In these improved dynamos, all soldered electrical connections in the armature were gold plated to keep internal resistance to a minimum. His second type of dynamo was called Jumbo, after P. T. Barnum's famous circus elephant. The first commercially manufactured Jumbo was shipped and exhibited at the Paris Electrical Exposition of 1881. The steam generator-dynamo unit weighed 27 tons. Edison's very popular exhibit there included a complete electrical distribution system with 500 of his 16-candle lamps. The Edison display easily took first honors and was largely responsible for the subsequent rapid growth of the Edison lighting system in continental Europe. Dynamos were manufactured by the Edison Machine Works, located originally at 104-106 Goerck Street, New York City. This was the first Edison company transferred to Schenectady, New York in December 1886.

Courtesy Edison National Historic Site,
West Orange, N. J.

Edison at age 34—a time of peak productivity in his innovative
career

In Edison's creation of a new system of electric lighting and
power distribution, numerous additional accessories besides
those already mentioned were developed and manufactured.
These included light and fuse sockets, switches from the small
snap type to large circuit breakers, ammeters, voltmeters, other
testing and control equipment, junction boxes, and light fixtures.

Sigmund Bergmann was one of Edison's workers at the bench
in Edison's first factory in Newark. When the move was made

to Menlo Park in 1876, Bergmann went to New York and opened a small shop of his own at 104 Wooster Street. His initial labor force consisted of one man and two boys. Edison began at once to entrust Bergmann with the manufacture of his various inventions, and by 1881 Bergmann had 50 employees. In 1882, when he took over the responsibility of manufacturing most of the electrical equipment Edison needed except the light itself, dynamos, and underground mains, Bergmann moved to a six-story building at 17th Street and Avenue B. Bergmann and Company, in which Edison had a one-third interest, manufactured literally hundreds of different electrical accessories. Bergmann returned to Germany in 1895, a wealthy man. He entered into electrical manufacturing there and at times employed more than 30,000 men in his works. Bergmann was one of several Edison employees who utilized their early experiences with the master to become highly successful business men later.

The area chosen for the first central lighting plant in New York City included the Stock Exchange and the great banking houses such as that of Drexel, Morgan and Company, the two most influential newspapers, the *Herald* and the *Times,* as well as small residential and factory areas. The location again showed Edison's daring approach to his new developments since these potential customers were in a position to make or break him from the standpoints of both publicity and financial backing.

The arduous task of laying the underground mains and feeder lines began in the fall of 1881. Edison had set up headquarters at 65 Broadway in the spring of that year, but he spent most of his time, day and night, supervising the installations and checking on the manufacturing operations in New York. He left the manufacture of lamps largely to his able colleague, Upton. Edison well realized that the junctions of the 20-foot underground conduits were critical; he allowed only Kruesi, head of the electric tube shop, Batchelor, or himself, to carry out this operation. Here the copper rods were coupled using flexible copper cables, any service wires were connected, and insulation was applied. On the first day of operation of the historic Pearl Street Station, September 4, 1882, 10½ miles of mains and 4½ miles of feeder lines had been installed, using approximately 180,000 lbs of copper. The system was brought into operation stepwise; initially, only one of the six dynamos was used, lighting an area having about 400 lamps. The success of the installation led to the complete laying of underground conduits by the end of 1882 in the one-square-mile area covered by the franchise. Additional areas of the installation were brought in as the operations were smoothed out. By October 1 1,284 lamps were in service; by the end of 1882 there were 231 customers with 3,477 lamps, and by December 1, 1883, there were 513 customers with 10,300 lamps served by 18 miles of mains and feeders. Individual customers increased their usage. As an example, within five months the *New York Times* had increased their number of lights five-fold to a total of 288. The first motor was connected to the system in 1884.

Edison's decision to place his conduits underground was not only a technical success but a farsighted move. Thus, in 1884 the New York State Legislature enacted a law requiring the underground placement of "all telegraph, telephonic, and electric light wires and cables used in any incorporated city of this state, having a population of five hundred thousand or over." However, the law was not enforced, and lower Manhattan continued to be shaded with a mass of electric, telegraph, burglar alarm, stock ticker, and telephone wires. Companies rarely shared poles, each erecting its own. The poles varied up to 150 feet in height, with as many as 12 cross-bars each holding 20 or more wires attached to one pole. The famous blizzard of 1888 brought a mass of overhead wires down on the stricken city and proved to be the last straw. Hugh Grant, New York City's new mayor, had promised enforcement of the 1884 state law if elected. Despite a court injunction obtained by Jay Gould of Western Union prohibiting the city to clear out the overhead wiring, 2,495 poles and 14,500,000 feet of wiring were removed in 1889, and the remaining were cleared by 1890. Small companies were forced out of business, and the electric companies which furnished current for arc lighting faced a serious problem because of its relatively high voltage.

Edison's decision to go underground also proved to be a wise one technically. Although many had claimed that an electrical conductor would be grounded if buried, regardless of how well it was insulated, Edison's underground tubes proved to be quite satisfactory, and by 1892 there were 172 miles in service in New York. Although electric cable came into use for underground wiring about this time, in 1907 Consolidated Edison Company of New York still had in service 54.3 miles of feeders and 185 miles of mains of the Edison type. None remain today; they have been superseded by cables in ducts (1).

Following the successful Pearl Street installation, the Edison system became widely accepted. About this time Edison needed to devote most of his time to business matters. This is evident from his patent record during this period. He applied for 24 patents in 1884 and 17 in 1885. The latter appear to have been mostly the continuation-in-part type, with 16 of the 17 dealing either with the telegraph or telephone. In contrast to these lean years, a total of 107 applications which subsequently resulted in patents had been filed in 1882, and these dealt nearly wholly with electrical inventions. Edison had transferred his headquarters from Menlo Park to 65 Broadway, New York, in February 1881, and by May of that year, the small staff remaining at the laboratory was moved to other Edison operations either in New York or abroad. Edison set up small laboratory quarters on the upper floor of the six-story

Lower Broadway in the late 1880's. Telegraph wires constituted most of the overhead array.

building which housed Bergmann and Company. His family had an apartment in New York, but the Menlo Park home was occupied during the summer months on a part-time basis. Following the death of his wife, Mary Stilwell Edison, in August 1884, Edison apparently abandoned any idea of ever returning to Menlo Park.*

By 1886 Edison tired of being a business tycoon and longed to get back to the laboratory. In early 1886 he married again and brought his bride, the former Mina Miller, to their new home in West Orange, New Jersey.** Soon construction of laboratory facilities began nearby. In 1889 he sold his interests in the various Edison manufacturing and illuminating companies to Henry Villard, a former railroad magnate, and his financial backers. Edison received over $1,000,000 in cash and approximately 10% of the stock of the new company. This appeared to have been a happy arrangement for Edison since his people continued to run the various affiliated companies, and additional capital was available for expanding the business.

However, Edison was not happy when in 1892 the Edison General Electric Company—the name of the combined company under Villard—was merged with the Thomson-Houston Electric Company to form the General Electric Company. Charles A. Coffin, former president of Thomson-Houston and the first president of General Electric, proceeded to purge the new company of Edison's name and most of his influence. Perhaps it was just as well since Edison had placed the Edison General Electric Company in a defensive position competition-wise by refusing to recognize the merits of alternating current (6). Nevertheless, the illuminating companies as a whole were unwilling to drop the magic word "Edison" from their company

---

* The first Mrs. Edison died of typhoid fever at the age of 29. There were three children by this marriage: Marion, Thomas Alva, Jr., and William Leslie.
** Mina Miller was the daughter of Lewis Miller of Akron, Ohio, a wealthy manufacturer of farm tools. After high school Mina had attended finishing school in Boston and then toured Europe. Miss Miller was 19 years of age at the time of her marriage, February 24, 1886.

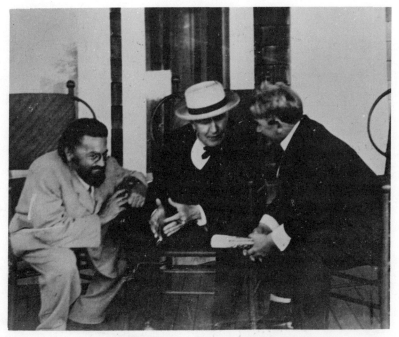

Courtesy Consolidated Edison Co. of New York

Edison with Charles Steinmetz of the General Electric Company and Arthur Williams of the Edison Electric Illuminating Company of New York

names, and most keep it to this day. Considerable difference of opinion exists as to whether Edison was or was not "frozen out" (*7, 8*). According to his children, Edison's only regrets concerning the merger were that he lost control of the lamp works, his first and favorite electrical manufacturing company, and that very few of his key people were retained in responsible positions (*9*).

Anyone familiar with the trials, headaches, and setbacks encountered in bringing any new technical development to fruition in the marketplace is impressed by his successes as Edison moved from numerous laboratory developments to commercial operations of his lighting and power systems. If one were to choose any characteristic of the man which made this possible, it would probably be his intimate knowledge of all phases of the system from the generation of the current to its final use by the customer. Most important, he did not hesitate to apply this know-how to the most humble task when it was necessary to solve a problem at hand. The following episodes illustrate this feature of his character.

In late 1882 an isolated electric plant was installed at the new Bijou Theatre in Boston. On opening night the Massachusetts governor and his staff in dress uniform were present as was Edison and several other Edison Electric Light Company officials in formal evening attire. During the second act the lights began to dim, and Edison, followed by Company President Johnson, dashed to the power plant which was located about 500 feet away in the basement of a printing plant. They found that the engineer and boilerman, struggling to repair a steam leak, had neglected the boiler. When Edison and Johnson did not return, other members of their party followed and discovered them, their tail coats and silk hats on a peg, busily shoveling coal into the firebox of the boiler. The show went on with good lighting maintained, but because of their soiled clothes, the two honored guests had to be excused from the formal speech-making ceremonies which followed.

The morning newspaper accounts following the opening of the Pearl Street Station were contradictory. The *World* described Edison dressed for the occasion in a "Prince Albert coat, starched white shirt, a white cravat, and with a white, high-crowned derby." Another newspaper described him as "disarrayed and collarless, with his white derby all stained with grease." Both descriptions were correct. Edison had dressed formally for the official ceremony in Morgan's office when the lights came on at 3 p.m., but soon afterwards he was in a manhole determining why a safety fuse had blown.

At the Southern Exposition held in Louisville, Kentucky, in

the summer of 1883, Edison's company exhibited his electric locomotive and a circuit of 4,600 lamps. When Edison visited the exhibit in August, the city planned a formal banquet in his honor, to begin on his arrival from New York. He reached the exposition hall on schedule but immediately disappeared into the basement where workmen were trying to repair the power plant of his exhibit. Edison was soon covered with soot and grease while food at the banquet grew cold. The *Louisville Courier* reported the following day that this instance illustrated that electricians and other workmen admired him so much because of "his attitude of caring nothing for what others thought of him, and paying no attention to frills and fancies" (*10*).

THE INVENTION OF the phonograph was not only unique but had several other unusual aspects. Edison's overall objective in life was to develop new things of a practical nature useful to mankind. Although he didn't place a very high priority on the need of man to be entertained, the phonograph found its biggest use in the entertainment field and not in the business office. Because of partial deafness, Edison never was able to have the full pleasure which the phonograph afforded to millions of others. Although earlier investigators had been able to produce speech-like sounds by mechanical means, as evidenced by mamma dolls and cuckoo clocks, apparently no one had ever conceived that sounds could be recorded, stored, and reproduced. Edison's first patent on the phonograph was granted without a single objection cited against it.

The reasoning which led to Edison's conception of the talking machine was based on his observations during experiments relating to recording telegraph messages and his experience with diaphragms in work relating to telephone transmitters and receivers. Following is Edison's own account of the invention of the phonograph (*1*):

I was experimenting on an automatic method of recording telegraph messages on a disk of paper laid on a revolving platen, exactly the same as the disk talking-machine of to-day. The platen had a spiral groove on its surface, like the disk. Over this was placed a circular disk of paper; an electromagnet with the embossing point connected to an arm travelled over the disk; and any signals given through the magnets were embossed on the disk of paper. If this disk was removed from the machine and put on a similar machine provided with a contact point, the embossed record would cause the signals to be repeated into another wire. The ordinary speed of telegraphic signals is thirty-five to forty words a minute; but with this machine several hundred words were possible.

From my experiments on the telephone I knew of the power of a diaphragm to take up sound vibrations, as I had made a little toy which, when you recited loudly in the funnel, would work a pawl* connected to the diaphragm; and this engaging a ratchet-wheel served to give continuous rotation to a pulley. This pulley was connected by a cord to a little paper toy representing a man sawing wood. Hence, if one shouted: "Mary had a little lamb," etc., the paper man would start sawing wood. I reached the conclusion that if I could record the movements of the diaphragm properly, I could cause such record to reproduce the original movements imparted to the diaphragm by the voice, and thus succeed in recording and reproducing the human voice.

Instead of using a disk I designed a little machine using a cylinder provided with grooves around the surface. Over this was to be placed tinfoil, which easily received and recorded the movements of the diaphragm. A sketch was made, and the piece-work price, $18, was marked on the sketch. I was in the habit of marking the price I would pay on each sketch. If the workman lost, I would pay his regular wages; if he made more than the wages, he kept it. The workman who got the sketch was John Kruesi. I didn't have much faith that it would work, expecting that I might possibly hear a word or so that would give hope of a future for the idea. Kruesi, when he had nearly finished it, asked what it was for. I told him I was going to record talking, and then have the machine talk back. He thought it absurd. However, it was finished, the foil was put on; I then shouted "Mary had a little lamb," etc. I adjusted the reproducer, and the machine reproduced it perfectly. I was

---

* A pivoted bar adapted to engage with the teeth of a ratchet wheel, or the like, so as to impart motion.—Ed.

never so taken aback in my life. Everybody was astonished. I was always afraid of things that worked the first time. Long experience proved that there were great drawbacks found generally before they could be got commercial; but here was something there was no doubt of.

The phonograph name is hardly the proper word to describe a machine for reproducing sounds from a record. The two syllables of phonograph would appear to have been borrowed from telephone and telegraph, basically meaning recorded voice. The term phonograph was coined not by Edison but by F. B. Fenby of Worcester, Massachusetts. In 1863 Fenby was granted a patent entitled "The Electro Magnet Phonograph" (2). This was actually an electrical device for recording the dots and dashes of the telegraph. Another mechanism which might be termed a forerunner of Edison's phonograph was the "phonautograph" of Leon Scott, a Frenchman. In a study of the vibrations of tuning forks in 1837, Scott built a recording device in which a pig's bristle attached to the center of a diaphragm of thin, stretched sheepskin, traced vibratory patterns upon lampblack-coated paper wrapped around a revolving cylinder. The bristle stylus left a visual record of the sound waves which struck the diaphragm. The cylinder was moved along by a screw as it rotated, much as in the later phonograph. However, Scott's stylus was lateral to the moving surface, whereas that of Edison's was vertical. Scott apparently made no attempt to reproduce sounds from his recordings, and it is highly unlikely that his recordings could be reproduced.

Edison made major contributions to the art of automatic telegraphy using both chemical and embossing methods. He developed devices whereby incoming messages could be stored and later played back to the listener at controlled speeds. This was done not only with the Morse dot-and-dash code, but he had some success in devising machines to record words. The latter, of course, is the actual practice today. Thus, it is not surprising that Edison conceived the desirability of storing telephone messages at the receiving end.

While working with Bell's telephone receiver Edison had observed that the vibrations of its diaphragm in tune with the voice were of considerable amplitude. During the summer of 1877, while attempting to repeat the indentations of telegraphic dots and dashes at a high speed from a tape of paraffin-coated paper, he noted "a light musical, rhythmical sound, resembling indistinct human speech." Edison's notebooks show sketches as early as July 18 which contain certain features of the subsequent phonograph. His experiments in the attempt to use paper for recording sound and reproducing it mechanically or electrically came to naught.

The actual date that Edison made the sketch of the grooved cylinder and accessories and gave it to Kruesi to construct has been a subject of considerable confusion in various books and magazine articles. Many have used a date of August 12, 1877. This has been proved incorrect, based on information obtained from Messrs. Norman R. Speiden and Harold S. Anderson in connection with their review of the original manuscript of this book (*see* Preface). The machine as discussed by Edison in the above citation was designed November 29 and tested December 6, 1877. The series of events which led to the establishment of these dates has all the earmarks of a good detective story (*3*).

On December 7 Edison and Batchelor demonstrated the machine in the office of the editor of *Scientific American*. The demonstration proved to be a sensation. The invention was featured in New York papers the next day and the subsequent issue of *Scientific American* of December 22. News reporters and the public flocked to Menlo Park to see and hear the "speaking phonograph." Edison was described by the press in such terms as the "New Jersey Columbus" and the "Wizard of Menlo Park." In Paris the directors of the International Exposition of 1878 accepted an Edison phonograph for display, even though the scientific jury could not determine the category of industrial arts to which it belonged.

Edison's basic patent application entitled "Improvement in Phonograph or Speaking Machines" was filed December 24, 1877 and issued as U.S. Patent 200,521 less than two months later, February 19, 1878. The fact that Edison used an old name—*i.e.*, phonograph—to describe his completely new machine would indicate that at the time he did not fully appreciate the uniqueness of his invention. Also, in the patent application, reference is made to two applications on file dealing with automatic telegraphy as well as comparisons with techniques for registering the Morse code. However, on the plus side, U.S. Patent 200,521 does disclose that the record upon the tin foil may be stereotyped "by means of the plaster-of-paris process" so that multiple copies could be made. It also discloses that a revolving plate could be used in the recording and reproduction as well as a cylinder.

In 1878 Edison filed an omnibus patent application entitled "Recording and Reproducing Sounds" in both the United States and England. It was filed on April 24, 1878 in the latter country and issued as British Patent 1,644 on August 6, 1878. This patent consists of 12 pages of specification, plus 67 diagrammatic sketches as figures. Its purpose was to disclose a broad range of possibilities for constructing all phases of the phonograph to prevent competitors from filing on these. For example, on page 2 of the patent Edison states, it is that the phonogram—*i.e.*, the indented material—"may be in the form of a disc, a sheet, an endless belt, a cylinder, a roller, or a belt, or strip . . ."

The corresponding U.S. application was not allowed, the Patent Office contending the English patent constituted prior publication. All details are not known by the author, but it is likely that Edison's lawyers erred in filing the British application first. In any case, it was a sad state of affairs in that Edison had disclosed a vast amount of technical information helpful to competitors, with no patent protection covering the same in the United States.

Edison was invited to demonstrate his phonograph before the National Academy of Sciences at its 1878 spring meeting in the

Edison with his host, Uriah Painter, and Charles Batchelor.
The phonograph shown is identical to the original except
that a flywheel has been added.

nation's capital. On the morning of April 18, accompanied by the faithful Charles Batchelor, Edison arrived, clothed in a new checkered suit. They were welcomed at the railroad station by Uriah Painter who informed them they had been invited that afternoon to the home of Gail Hamilton, a noted Washington hostess, to meet a group of congressmen and members of the diplomatic corps.

The group first visited the Brady Studio where Edison's picture was taken with the phonograph, as well as one including Painter and Batchelor. They then called on Joseph Henry, President of the National Academy, to pay their respects, and went on to Miss Hamilton's in the afternoon. Edison appeared before the Academy in the evening at the Smithsonian Institution. His demonstration was followed by an address by Professor George F. Barker, of the University of Pennsylvania, describing the successful transmission of speech 140 miles by telephone using the new Edison carbon transmitter and induction coil. Following Barker's speech, he and Edison were given a standing ovation.

President Hayes heard the news of the phonograph demonstrations, and the Edison party was invited to come to the White House following the meeting of the Academy. They arrived about 11 p.m., and Edison and Batchelor were returned to their hotel by presidential coach after 3 a.m. The President had been so impressed that he induced his wife, who had retired, to get up and join the group. She did, accompanied by a group of other ladies. Some chemists and other scientists worry about their professional status in relation to doctors and lawyers. After his eventful day in Washington, Edison should have had no qualms in this respect.

Although Edison's objective in the phonograph development was to use it as a recorder of telephone messages, the unexpectedly high quality of reproductions obtained early convinced him that the instrument had other potential uses. In a published article on the phonograph in June 1878, Edison listed the following under the heading of "probabilities" (*4*):

> Dictation without the aid of a stenographer
> Phonographic books for the blind
> Teaching of elocution
> Musical and entertainment records
> Teaching languages
> Family record of voices
> Annunciators for clocks, etc.
> Preserve voices of the great
> An auxiliary in use of telephone

Edison often has been quoted as saying that the phonograph was his favorite invention. Early in 1878 he described it as "my baby, and I expect it to grow up and be a big feller and support me in my old age." As we have seen from Chapters 2 and 3, the parent was too busy in the electrical field to give the baby much attention during the first 10 years following its birth.

The competitive business aspects of the manufacture and sale of phonograph machines and records had a major effect on the technological developments in this field by Edison and others. Following the epic invention in 1877, Edison gave the phonograph low priority in his research and development program. The telegraph and telephone were considered much more important at the time, and starting in the fall of 1878, work on the electric light overshadowed all else.

About 600 of the tin-foil machines were manufactured. The bulk of them were made by Sigmund Bergmann at his Wooster Street shop, although some were made in the shop at the Menlo Park laboratories and several small local shops. The cylinder of hollow brass was mounted upon a shaft, at one end of which was a crank for turning and at the other a balance wheel. As shown in the photograph (p. 103), the shaft, threaded at one end, was mounted on two iron uprights. The mouthpiece of gutta-percha was attached to a thin metal diaphragm, against the center of which was an embossing point held by a spring attached to the rim of the mouthpiece. A continuous helical groove encircled the cylinder, and a sheet of tin foil was wrapped around it. For recording, the embossing point was pressed against the foil, and as one spoke into the mouthpiece, the cylinder was rotated with the screw crank. About one word per revolution was recorded. To reproduce, the cylinder was turned back to its original position, the stylus point fitted into the indented grooves of the tin foil, a horn or earphones attached to the mouthpiece, and the cylinder was again rotated.

A group of venture capitalists set up the Edison Speaking Phonograph Company, and the machines were leased by various exhibitors for showing around the country. Edison received a down payment of $10,000 and 20% of gate receipts. The business thrived for about a year, but after most of the country's population had seen or heard about the novelty recordings, audiences at the music-box parlors dwindled. The tin-foil phonograph was a crude device, crying for improvements. Rather than abandon the invention as he pursued "bigger game," it is unfortunate that Edison did not turn the development over to one of his promising men, such as Edward G. Acheson. Edison hired Acheson in 1880 but the latter became dissatisfied when he was not allowed to experiment with any of his own ideas. Edison tended to put such people on manufacturing jobs. As a result, Acheson left Edison's employ and became famous in his own right as the inventor of silicon carbide (Carborundum) and various graphite emulsions.

In 1880 Alexander Graham Bell, inventor of the telephone, won the Volta Award granted by the French Academy of Sciences. The monetary award was 50,000 francs (about $10,000) —a very generous figure even by today's standards. Bell used most of the money to organize the Volta Laboratory Association and opened a research laboratory in Washington, D.C. Very early, work on the phonograph was emphasized. Bell was joined by his cousin, Dr. Chichester Bell, a chemist on the faculty of the University of London, who had come to the United States on a leave of absence. A second associate was Charles Sumner Tainter, an optical instrument maker. The Bells and Tainter were granted five patents relating to the phonograph on May 4, 1886. The only one to become important in later patent litigations was 341,214. The inventors utilized a wax-coated cardboard cylinder for the recording surface. Their point of novelty as allowed by the Patent Office, and later sustained by the courts, was that the recording stylus *incised* the wax surface rather than *indented* as in the case of a metallic foil. Because of physical differences of the two classes of materials, a recording stylus that would only indent tin foil would automatically cut a groove in a wax surface. This was certainly a matter of semantics and not science.

Bell and his associates formed the Volta Graphophone Company in 1886. They sought to combine forces with Edison, but he would have none of it. However, the Graphophone development shocked him into action, resuming long working days in the laboratory. During 1888–89 Edison filed many patents dealing with the phonograph. A wax cylinder record of about $\frac{1}{4}$-inch wall thickness was developed. As in the case of the original tin-foil phonograph, as much emphasis was on recordings by the customer as was on reproductions. The wax cylinders were made sufficiently thick so that many recordings were possible; each previous recording could be removed by a cutting tool provided with the machine. The wax cylinder slid onto a tapered metal cylinder on a shaft which was powered by an electric motor driven by a primary or storage battery. Although designed primarily for office dictation, the phonograph found its main use in the entertainment field. The advent of the coin-in-the-slot phonographs began in earnest in 1892.

As president of the Edison Phonograph Company, formed in 1888, Edison granted a contract for exclusive sales of his phonograph to Ezra Gilliland, a friend since their fellow telegrapher days. Gilliland, in turn, arranged for Edison to sell the stock of Edison Phonograph to the North American Phonograph Company, headed by Jesse Lippincott. The latter company was to control subsequent patents on the phonograph obtained by Edison and to have exclusive sales rights on machines and records. Manufacturing and development were to be carried out by the Edison Phonograph Works. Certainly Edison was trying as much as possible to stay out of the business end of the venture. This was 1888, and he was still very involved in all phases of incandescent electric lighting and had only recently occupied his expanded research quarters at West Orange.

Lippincott also contracted with the American Graphophone Company to market their machine. The whole venture was poorly handled; Lippincott's health failed, and he died in 1892. American Graphophone Company brought suit against Edison for alleged violation of their patents. Angered by what he considered bungling by his advisors, Edison as chief creditor of the North American Phonograph Company forced the company into friendly receivership in order to regain control of his own patents and sales.

By February 1896 Edison had regained the rights to his own patents but suffered the humiliation of being prohibited by the courts to sell phonographs in the United States for three years. Fortunately, export sales were good during that period. Shortly after the turn of the century, the phonographic "craze" arrived, and musical records were in great demand. In 1896 Edison factories in West Orange began manufacturing a spring-powered phonograph expressly for home use. Cylinder records for these machines played for two minutes and cost 50¢ each.

In 1908 Edison began marketing cylinders which played for four minutes. In 1911 Edison's National Phonograph Company and his other separate corporations were merged into one corporation, Thomas A. Edison, Incorporated. In 1913 the Edison Diamond Disc Phonograph was introduced. It was this machine which was used in the famous Tone-Test Recitals to demonstrate that "the Edison Diamond Disc's recreation of music cannot be distinguished from the original." In 1914 Edison's annual phonograph business surpassed $7,000,000.

Beginning in 1923 the phonographic business began to ebb seriously as radio came into general use. The creator of the mechanical phonograph refused to introduce electronic recording or to market a volume-controlled electronic phonograph. He detested the radio, saying it gave no choice to the listener. Finally, in 1925 on urging from his two sons, Charles and Theodore, who were then active in the company business, work on an electronic phonograph was begun. Such a product was introduced in 1928, but because of the head start of competitors in the new technology of recording and reproducing and the general business depression which began in the fall of 1929, the Edison Company on November 1, 1929 announced it would discontinue manufacture of all phonographs and records except those for business purposes. Originally Edison's business phonograph had been called by various names including the Ediphone but later was given the trademark name of Voicewriter. In 1957 Thomas A. Edison, Incorporated, merged with McGraw-Electric Company to become a part of the McGraw-Edison Company.

The American Graphophone Company had been unable to make a satisfactory machine until 1893 when it adopted several features of the Edison machine, such as tapered mandrels, solid wax blanks, and sapphire recording and reproducing styluses. Because of the joint arrangement with North American Phonograph, they were able to do this. A series of lawsuits and price wars between the two companies continued past the turn of the century. Ironically, in October 1897 Edison had to mortgage the Edison Phonograph Works for $300,000 because of heavy expense in connection with his iron concentration demonstration plant (*see* Chapter 5). At the same time, American Graphophone was highly prosperous and was advertising the Graphophone as a perfected version of the Edison phonograph.

The Columbia Phonograph Company was one of the local companies under the North American Company which became independent. This company, marketing primarily in the Washington, D.C. area, took aggressive steps in 1893 in an attempt to gain a dominant position in the phonograph field. Columbia gained control of the American Graphophone Company in May 1893 and thus ownership of the Bell-Tainter patents. For all practical purposes, the two companies were one. In 1906 the merged companies reorganized under the name of Columbia Graphophone Company. Because of a stock promotion scheme in 1918, plus the advent of radio in the early 1920's, this company went bankrupt in 1923. A segment of the company was reorganized as the Columbia Phonograph Company by the British affiliate, Columbia Phonograph Company, Limited, of London. The American company was instrumental in the reorganization of Independent Broadcasters, Incorporated, into the present Columbia Broadcasting System (CBS). Columbia Phonograph became bankrupt during the prolonged depression in the early 1930's but was rescued by its robust child as Columbia Records, Incorporated and is today a wholly-owned CBS subsidiary. The Graphophone Company Limited of Middlesex, England, continues under that name.

The most successful business venture involving the phonograph was that initiated by Emile Berliner, a young Boston scientist. Berliner had worked earlier in the telephone field and developed a phonograph known as the Gramophone which used disc records. From the start the Gramophone was designed to be used for reproductions only and was sold primarily to provide "canned music" for the home. Multiple records were made from a single master which was a zinc disc with a continuous spiral groove. The top surface of this disc was covered with a wax, and the recording stylus cut through the wax and contacted the grooves laterally, rather than vertically as practiced by Edison. The zinc disc surface was then treated with chromic acid which attacked only the metal surface exposed by the recording stylus. This served to accentuate the varying indentations of the stylus. The wax was then removed by a solvent, a "negative" was made by electrodeposition of copper or other metal, and this was used to press out recordings on hard rubber discs. This technique of making multiple recordings from a single record was disclosed by Edison, but the technique was much easier with a flat disc recording than with a cylinder. Edison favored cylinders because of the uniform speed at the point of contact with the stylus during both recording and reproducing.

Other firsts practiced by Berliner were the use of a horn rather than the earlier use of earphones, a floating stylus which followed the grooves of the disc, and the use of a spring motor to turn the record.

Founded on the basis of Berliner's Gramophone patents, the U.S. Gramophone Company was organized in 1893 and the Berliner Gramophone Company the following year. The original Gramophone machines were hand cranked. Eldridge R. Johnson, who operated a machine shop in Camden, New Jersey, was engaged to manufacture spring motors for the Gramophones and later to make the entire machine. Johnson became a partner of Berliner, and the Victor Talking Machine Company was organized in 1901. With such innovations as the Victrola with its horn concealed and the famous "His Master's Voice" trademark, Victor became a highly successful manufacturer of phonographs and musical disc records. When sales declined in the 1920's, an electronic amplified Orthophonic Victrola was developed and marketed in 1926. This was the same year that the Radio Corporation of America organized the National Broadcasting Company, and in 1929 RCA acquired the Victor Talking Machine Company.

Although Edison's phonographic business was probably his best source of income over the years, the whole industry was plagued for years with lawsuits and unscrupulous marketing and advertising practices. It is almost unbelievable that the Patent Office and the courts did not consider the original Edison patent as the one to dominate all speaking machines. This would have given Edison overall control of the industry until 1895, subject, of course, to improvement patents by himself and others. Instead, we find his patent restricted to the tin-foil concept and the inventor of one of the most original of all inventions in a secondary patent position during the 1890–1905 period, when this infant industry was being established and giant strides made.

Although the phonograph was invented at the Menlo Park laboratories, research on all its phases continued at Edison's West Orange laboratories over a period of about 40 years. This was Edison's longest sustained research program in any one field, and the West Orange laboratories played a significant role in his overall program.

When Edison moved to New York in 1881 to concentrate on manufacturing materials for electric lighting and establishing the Pearl Street Station, he set up a small laboratory on the top floor of the offices of the Edison Electric Light Company at 65 Fifth Avenue. After Bergmann and Company moved from 104-108 Wooster Street to 17th Street and Avenue B, in October 1882, Edison moved his laboratory to the 6th floor of that building. John Ott, one of Edison's early and most trusted assistants, resumed work on the phonograph at the Bergmann site in late 1886. Shortly after the move to the spacious new home of "Glenmont" in Llewellyn Park, West Orange, Edison purchased a tract of rural land about one-half mile from his home. This valley was to be the site of his new laboratories, with ample space for factories to manufacture the new products which would be developed. Edison's own words in a letter to J. Hood Wright, one of J. P. Morgan's partners, describe his enthusiasm and high expectations for his new venture (5):

Mr. Wright, my laboratory will soon be completed—the dimensions are one building 250′ long, 50′ wide, and 3 stories. Four other buildings 25 x 100 one story high, all of brick. I will have the best equipped and largest laboratory extant, and the facilities incomparably superior to any other for rapid and cheap development of an invention, and working it up into commercial shape with models, patterns and special machinery. In fact, there is no similar institution in existence. We do our own castings, forgings. Can build anything from a ladies watch to a locomotive. The machine shop is sufficiently large to employ 50 men, and 30 men can be worked in the other parts of the works. Inventions that formerly took months and cost a large sum can now be done in 2 or 3 days with very small expense, as I shall carry a stock of almost every size and with the latest machinery a man will produce 10 times as much as in a laboratory which has but little material not of a size, delays of days waiting for castings and machinery not universal or modern.

You are aware from your own acquaintance with me that I do not fly any financial kites, or speculate, and that the works which I control are well managed. In the early days of the shops it was necessary that I should largely manage them, first because the art had to be created, second, because I could get no men who were competent in such a new business, but as soon as it was possible, I put other persons in charge. I am perfectly well aware of the fact that my place is in the Laboratory, but I think you will admit that I know how a shop should be managed and also how to select men to manage them. With this prelude, I will get to business—my ambition is to build up a great Industrial Works in the Orange Valley starting in a small way & gradually working up; the Laboratory supplying the perfected

113

This drawing of the Edison Laboratories and the phonograph works appeared in the May 1903 issue of *Phonograph Monthly.*

invention models, patterns and fitting up necessary special machinery in the factory for each invention. My plan contemplates to working of only that class of inventions which require but small investment for each and of a highly profitable nature, and also of that character that the articles are only sold to jobbers, dealers, etc.—No cumbersome inventions like the Electric Light. Such a works in time would be running on 30 or 40 special things of so diversified nature that the average profit would scarcely ever be varied by competition, etc.

In this letter to Wright, Edison went into more detail about how potential companies for exploiting the laboratory developments would be set up and ended with the following question: "Now, Mr. Wright, do you think this practicable; if so, can you help me along with it?" There is no record as to Mr. Wright's reaction to the letter, but it is likely that he and his associates at Drexel, Morgan, & Company would have highly preferred that Edison concentrate on making their joint electric lighting business more profitable rather than striking out on new ventures. In any case, after Morgan and his associates promoted the merger of the Edison manufacturing companies with those of Thomson-Houston, Edison had had enough of Wall Street bankers, and he went it alone in financing his subsequent manufacturing operations.

The West Orange laboratories were completed in late 1887, aggregating in floor space about 10 times the size of the Menlo Park installation. At that time, the West Orange establishment was the most elaborate industrial laboratory in the world; in physical plant and organization it set the pattern for the laboratories of large corporations which began to appear after the turn of the century. Chemistry was emphasized here much more than it had been at Menlo Park. The functions of the four one-story laboratories were originally as follows, numbering from the main gate as shown in opposite sketch: (1) electrical and physics building, (2) chemical laboratory, (3) chemical storage and pattern shop, and (4) metallurgy.

The main building consisted of five chief divisions—the library, offices, machine shops, stockrooms, and experimental rooms. One of the latter was the "Music Room" on the third floor, with a huge music library. For about 40 years it was the headquarters for phonograph experiments. Commercial recordings of musical wax cylinders started there in 1889, and supervision and control of selection, approval, matching, and releasing of phonograph recordings were carried on, with Edison's active participation.

The West Orange laboratories, now the Edison National Historic Site, National Park Service of the U. S. Department of Interior, can be seen by the visitor much as they were in Edison's day. In the library, a large room with a 30-foot high ceiling and two tiers of galleries rising from its main floor, is a vast collection of approximately 10,000 books and bound periodicals covering a range of subjects. William H. Meadowcroft, Edison's associate and confidant from 1881 until the inventor's death, describes the contents of the library as it was in 1911 (6):

The library is a spacious room about forty by thirty-five feet. Around the sides of the room are two tiers of gallery. The main floor and the galleries are divided into alcoves, in which, on the main floor, are many thousands of books. In the galleries are still more books and periodicals of all kinds, also cabinets and shelves containing mineralogical and geological specimens and thousands of samples of ores and minerals from all parts of the world. In a corner of one of the galleries may be seen a large number of magazines relating to electricity, chemistry, engineering, mechanics, building, cement, building materials, drugs, water and gas power, automobiles, railroads, aeronautics, philosophy, hygiene, physics, telegraphy, mining, metallurgy, metals, music, and other subjects; also theatrical weeklies, as well as the proceedings and transactions of various learned and technical societies. All of these form part of Mr. Edison's current reading. At one end of the main floor of the library, which is handsomely and comfortably furnished, is Mr. Edison's desk, at which he may usually be seen for a while in the early morning hours looking over his mail.

An interior view of the library as it appears today

A glance at the book-shelves affords a revelation of the subjects in which Edison is interested, for the titles of the volumes include astronomy, botany, chemistry, dynamics, electricity, engineering, forestry, geology, geography, mechanics, mining, medicine, metallurgy, magnetism, philosophy, psychology, physics, steam, steam-engines, telegraphy, telephony, and many others. These are not all of Edison's books by any means, for he has another big library in his house on the hill.

The stockroom was said to contain over 8,000 items, including samples of all chemicals available from reagent suppliers, a great variety of plant and animal products, also minerals, fibers, tar, rosin, pitch, asphalt, graphite, and asbestos of various grades. All necessary items were stocked for the machine shop and for

117

woodworking, including all the known metals in the form of sheets, rods, and tubes. Based on materials available at that time, the stockroom had an inventory comparable with those of the multimillion dollar research centers of today.

Although Edison was often short of capital for his business operations, he was far from skimpy when it came to spending money for research and development. He once told William H. Mason, superintendent of his cement works and later the inventor of Masonite, "You must experiment all the time; if you don't the other fellow will and then he will get ahead of you."

Thus Edison had the facilities and no doubt the sincere desire to continue a vigorous research program on the phonograph. One may wonder why he permitted others to get ahead of him in several obvious improvements. Although the use of discs rather than cylinders for records as well as a spring motor for turning the records were disclosed in his early patents, he let others pioneer these important innovations in the marketplace. The answer was that Edison was always involved in a variety of projects, and the phonograph at times did not appear to be very high on his priority list. For example, A. O. Tate, Edison's private secretary in the early nineties, reported that in 1890 he was carrying cost records of 72 separate research projects in progress at the West Orange laboratories.

The manufacture of phonographic records introduced the first large scale fabrication of plastics. When one broadly defines a plastic as an organic material capable of being molded or otherwise formed into desired shapes, this includes not only polymeric materials but also resins, gums, waxes, solid hydrocarbons, asphalt, solid fatty acids and their salts, and the like. Edison and others used materials from each of these classes of chemicals to fabricate cylindrical and disc records.

Cellulose is a polymer of glucose (a simple sugar) molecules with a molecular weight the order of 500,000. Braconnot of France prepared cellulose nitrate and demonstrated its plastic properties. In 1869 the Hyatt brothers of New Jersey prepared celluloid by mixing camphor with cellulose nitrate to form thermoplastic moldable compositions. Celluloid is considered to be the first synthetic plastic. Hard rubber, also called ebonite or vulcanite, is formed by the vulcanization of natural rubber with 35–40% of sulfur, based on the final product. Unlike usual rubber vulcanizates containing only a few percent sulfur, hard rubber has no elasticity and is somewhat thermoplastic in nature. Hard rubber was known before celluloid was developed, and some may claim that it was the first synthetic plastic.

Courtesy Edison National Historic Site,
West Orange, N. J.

Interior view of the chemical laboratory

Resins are a large class of plastic-like materials. Amber and copal were fossil resins well known to the early applied chemists. Present-day synthetic resins such as unsaturated polyester, epoxy, and phenolic resins represent the more strict concept of a resin in that they are employed as liquid or low melting amorphous solids, but in the process of fabricating desired shapes, they are converted to hard, tough thermoset plastics. The conversion may occur catalytically or simply by condensation with itself or another substance. An important natural resin is shellac, which is the purified product from the secretion of an insect which is parasitic on selected trees and bushes in India, Burma, and Thailand. Shellac consists of various organic derivatives of polyhydroxy aliphatic acids. Although it is thermoplastic, shellac hardens when it is heated above 200°F for any length of time because of intermolecular reaction of the hydroxyl groups; thus, it becomes thermosetting, much as a phenolic resin. Rosin, one of the constituents in exudates from pine trees, was an important resin in Edison's time. It is primarily a mixture of resinous acids. As a thin film it will harden with time as a result of oxidation by atmospheric oxygen.

The chemist considers a true wax to be an ester of a high molecular weight fatty acid and a high molecular weight alcohol. Natural waxes are of both animal and vegetable origin. Those available to the early fabricators of phonograph records were beeswax, montan wax, carnauba wax, and spermaceti. The latter is available from the whale and is cetyl palmitate— *i.e.*, the combination of a C_{16} alcohol and a C_{16} fatty acid.

Mineral waxes are not true waxes but were called such because their physical properties resembled beeswax, the "grandfather" of all waxes. Mineral waxes include paraffin which is obtained from petroleum, ozocerite, a wax-like fossil found along with certain bituminous and coal shale deposits, and ceresin, a purified form of ozocerite. Mineral waxes are mixtures of hydrocarbons. Asphalt can also be considered a mineral wax, but in addition to hydrocarbons it contains many other complex bodies, including nitrogen and sulfur compounds. In the late 19th century, most asphalt of commerce came from asphalt lakes rather than residua from the distillation of petroleum. An asphalt lake in Trinidad, occupying about 100 acres and of uniform composition to a depth of 285 feet, was the primary source for the United States. As mined, this asphalt contains about 29% water and gas, and about 27% very finely divided silica and clay.

Gums have been used by mankind for thousands of years. The term "gum" was originally applied to plant exudates which had oozed from plant barks and hardened upon exposure; thus, the term "gum rubber" for raw natural rubber. However, today the descriptive term of gums is usually confined to water-soluble materials such as gelatin, gum arabic, and gum tragacanth. Such materials are proteins or sugar-type substances.

The Bells and Tainter used ozocerite-coated cardboard for records in their Graphophone development. Edison, on the other hand, developed a new class of plastic materials based primarily on blends of solid fatty acids and their salts. Such compositions were used in his solid cylindrical records starting in the late 1880's and continuing up to September 1912.

Stearic acid is a straight-chain C_{18} fatty acid, melting point, 157°F. Commercial grade stearic acid was then, as now, an approximate 50/50 blend with palmitic acid (C_{16}), obtained from the saponification of beef tallow. Since the two have nearly identical properties, it is economically unfeasible to separate them. Commercial stearic acid is less than 0.05% soluble in water at ordinary temperature, and in properties it resembles waxes more than a low molecular weight fatty acid such as acetic. However, inorganic salts of stearic acid can be prepared, that of sodium being used in soaps. Soaps contain sodium salts of other fatty acids such as that of oleic acid, an unsaturated liquid C_{18} acid, in order to obtain the desired water solubility characteristics. Edison not only used stearic acid and its sodium salt for preparing "wax" records, but he also studied in detail the preparation and use of various other metallic stearates, such as those of calcium, strontium, magnesium, lead, and aluminum. These are more gelatinous and less affected by moisture than that of sodium.

The improved Edison phonograph of 1888, using a solid wax-type cylinder about ¼-inch thick, was introduced first as an instrument of communication and as a business machine for dictation. The records would be used many times, which meant that the cylinder had to be relatively soft so that an old recording could be shaved off prior to reuse.

In 1896 the following wax composition was being used:

Stearic acid	48.0%
Sodium stearate	20.2
Aluminum stearate	11.3
Ceresin	20.5

The blend was prepared in approximately 900-lb batches. The best grade of stearic acid was melted in a lead-lined iron kettle and filtered through canvas into a copper kettle lined with silver. Crystalline sodium carbonate, $Na_2CO_3 \cdot 10\ H_2O$, was added at 260°F, and the temperature allowed to reach 280°F. When foaming had subsided, aluminum stearate and ceresin were added. The mass was heated to 440°–480°F until all foaming ceased. The temperature was then lowered to about 308°F and the molten wax was strained through three thicknesses of finely woven muslin and poured into pans to cool. The congealing point was to be about 260°F. If lower than this, the proportion of soda was increased in the subsequent batch and decreased if it was above 260.

High quality chemicals and expensive corrosion-resistant equipment were used in these operations. Filtrations removed any incompatible material. The fact that the sodium stearate was prepared from high grade stearic acid and soda crystals, rather than purchased in the form of the sodium soap, indicates the insistence on product quality rather than on lower cost. Originally the wax cylinder was prepared by dipping cold metal mandrels into the molten wax blend. The mandrels were ta-

Courtesy Edison National Historic Site, West Orange, N. J.

Edison in his chemical laboratory (1890)

pered, facilitating their removal from the solidified wax tubes. The cylinders were then cut to proper length, and the exterior was turned down to the specified uniform diameter on a lathe.

A tapered metal cylinder was used in the phonograph for mounting the cylinders.

Many problems were encountered with the wax compositions. Those records which were used for hundreds of recordings required a surface which was soft enough to record well but hard enough to reproduce. Unfortunately, minute cracks tended to form when the recording was shaved off before the surface was reused. Another problem involved temperature extremes. Home phonographs were usually kept in the parlor which was heated in the winter only on Sundays and special occasions. Thus, the wax records were subjected to temperatures as low as 0°F, where they became brittle and tended to crack, and temperatures as high as 100°F, where they tended to soften.

It is well known today that fiber-reinforced plastics, such as those containing glass filaments, have better dimensional stability with change in temperature and are stronger than unreinforced castings. Edison utilized this principle in wrapping thread, cord, paper, or other filament matter around the mandrel before impregnating and coating with his wax-like blend (7). Thus, the inner layer of the cylinder would be fiber reinforced.

Laminated cylinders were also made using a hard, tough material for the inner layer and the relatively pliable soap–fatty acid composition to receive the recording. One combination was a purified Trinidad asphalt containing a small amount of carnauba wax as the inner layer and the usual wax-like composition on the outer (8). The lower diameter asphalt cylinder was first formed by dipping or molding, followed by molding of the outer layer using a split mold and a "squirting press." Another technique was to mold the two cylinders separately, machine each to proper dimensions, heat the larger one, and slip it over the other; contraction on cooling resulted in bonding. Hard rubber, shellac-containing fillers and plaster of paris as possible materials for the inner layer also were disclosed in

Edison's patents. He recognized that the compositions making up the two layers should have comparable coefficients of thermal expansion in order to maintain good adhesion bonding. Adhesion must have been a problem since there are patent disclosures relative to the use of a thin third layer of paraffin, rubber, or the like, between the other two layers to facilitate bonding.

Warpage of the hot cylinders following their removal from the mandrel or mold also was a problem. Although the outer surface of the cylinder could be machined to uniform dimensions, the inner dimensions might change so that the cylinder would not fit the carrier of the phonograph. Spiral or parallel ribs as a part of the inner surface of the cylinders were developed to decrease the tendency to warp and to allow easy reaming of the cylinder to proper inside diameter if warpage did occur (*9*).

In the original phonograph patent Edison states that multiple copies of a given record can be made from a plaster-of-paris imprint of the indented tin foil. Thus, from the beginning it was assumed that a sound record could be repeated in multiple by making a negative of the "hills and valleys" of the recording and stamping or impressing this on a plastic surface. The reproduction of sound recording by this technique is indeed ingenious, when one considers that the recording grooves were about 0.001 inch deep and the differences in depth of the various recorded vibrations almost infinitesimal.

Berliner was the first to market records stamped from negatives of the master record. This was relatively easy since he used disc records exclusively. Celluloid was initially used by a licensee in Germany to receive the impressions, but hard rubber was used from the start in Berliner's manufacture in the United States, starting in 1893. Because of porosity problems, rejects were high, and grooves in the hard rubber relaxed and tended to disappear with age. In 1897 Berliner switched to a shellac composition containing inorganic fillers marketed by the Durinoid Company of Newark, New Jersey. Shellac records were employed by Victor and others until the 1940's when man-made vinyl plastics replaced all natural organic materials used in records. Edison never used shellac as a primary ingredient in either his cylindrical or disc records.

With cylinder records, the preparation of metal negatives and positive impressions from them was a much more formidable task than that faced by Berliner. Edison developed two techniques for preparing the cylindrical negative. One was to deposit a fraction of a mil of metal on the master recording using a vacuum deposition technique. Gold was preferred. Thin foils of gold were placed on either side of the recording, and the cylinder was rotated while subjecting the two foils to a high tension electric discharge under vacuum. The thin deposit obtained made the recording surface conductive so that the metal layer could be built up by electroplating, using such metals as copper or nickel. The cylindrical master record was removed from its metallic negative by dissolving, melting, or crushing after cooling to embrittle it.

Another approach was to make the surface of the recorded master conductive by coating it with ground graphite and electroplating it directly. However, initially this technique gave inferior results, and the gold-coating method was used until after the close of World War I. Paul Kasakove, one of Edison's young chemists, was instrumental in determining that if the graphite were finely enough divided and pure enough, it could be used effectively. This development came so late that it was used principally with discs.

Edison's original process for preparing recordings from such a metal negative was to cut it into halves and use them as a matched-metal mold for impressing wax cylinders under heat and pressure (10). This technique tended to result in recordings which were imperfect at seams where the two halves came together. This approach was replaced by using the negative cylinder whole, placing a slightly undersized wax blank within it, and then forcing the blank into the interstices of the negative at elevated temperature. The recorded cylinder was then cooled and removed. The fact that plastics have a much higher degree of contraction when cooled than do metals made possible the removal of wax cylinders from the metal negatives by cold shrinkage. Another technique was to build up a layer within the negative with molten wax which was subsequently removed after cold shrinkage. The inside of the cylinder so formed was then machined to proper size.

For disc records manufactured in volume, it was necessary to make so-called submasters, from which negatives were made to be used in the large volume molding operations. As many as 100 subnegatives might be made of a record which sold in hundreds of thousand units. In 1928 the procedure of such a duplication operation was as follows (*12*). Finely divided natural graphite (Dixon #97) was refined to at least 99% purity by treatment with hot aqueous sodium hydroxide, followed by water washing and drying. The graphite was applied to the master record with a silk-bristle brush while the record rotated at high speed. The master record so treated was fixed on a hard-rubber rotator and placed in a copper-plating bath. Plating was begun at a current density of 0.5 ampere which was increased slowly to 12 amperes over a period of four hours. The amperage was then increased to 50 and so maintained for 16 hours. Then the copper negative was removed, and the recorded surface was polished by a goat-bristle brush. A series of nickel-faced-on-copper submasters were then made by electroplating the master negative. The desired number of subnegatives for use in manufacturing records for sale were prepared from the submasters and were also nickel faced with copper backing.

When making cylindrical master and duplicate records, Edison encountered many of the problems met 60 years or so later by plastics fabricators. His solutions to such problems as removal of wax cylinders molded on metal mandrels, strengthening of the structure by design, and avoidance of incipient cracking and embrittlement of the structure were similar to those which have been used in the manufacture of fiberglass and other types of reinforced plastic pipe. To obtain grooves of proper dimension and spacing, he had to allow for shrinkage following the hot molding operation. Considering there were up to 200 grooves per inch and that these had to match precisely the threads moving the stylus, this was a design problem of no mean magnitude even by present-day standards.

Sound records were at first duplicated by recording the reproduced sound of the master record on other record blanks. When Edison used this method, the original was recorded on a relatively large diameter cylinder which allowed a higher circumferential speed, resulting in better recording and reproduction of higher frequencies. The 5-inch wax cylinder, used as a master from 1898 until the molding process took over, was also sold to the public for use on a special "Concert" model phonograph.

After it became apparent in the early 1890's that the main interest of the public in the phonograph was for musical records to be played hundreds of times, Edison strove to decrease the wear of the sound track caused in reproduction. Use of the usual steel needles and cactus thorns as styluses not only caused excessive wear, but the steel needles were corroded by the free stearic acid in the wax-like composition. In May 1890 Edison applied for a patent dealing with the use of hard crystals such as sapphire, quartz, and diamond for sound reproduction and recording. Sapphire is one of the grades of the mineral corundum, a crystalline form of aluminum oxide, and one of the hardest substances known. Edison first used sapphires as tiny rods, hemispherical at the ends, but in 1902 he introduced the "sapphire button" which was a doorknob-shaped stylus only slightly larger than the width of the record grooves. Such a stylus made it possible to play cylindrical records of the relatively soft soap–stearic acid composition with improved frequency response and less wear.

Although Edison's wax-type records were generally accepted as having acoustical properties superior to those made from harder materials such as shellac, the metal–soap compositions had serious shortcomings for longer playing records where the grooves had to be closer together. In 1908 Edison introduced his four-minute Amberol records which had 200 grooves per inch as compared with 100 for his previous two-minute records. A harder soap compound was used for the Amberols, but its greater brittleness led to more frequent damage in playing and handling. In 1912 the Blue Amberol was introduced which was a celluloid composition around a core of plaster of paris. Edison was forced to license the use of celluloid from the Lambert Record Company because of a patent issued to them in 1900 although he had experimented with celluloid previously, and in the early 1890's Berliner had actually marketed disc records made of celluloid. To play the celluloid four-minute records, Edison developed a diamond stylus which used a heavier floating weight.

From 1913 to 1916 Edison carried out an intensive research program on the phonograph to make the quality of musical reproductions comparable with the originals. He felt that the recording of classical music was in a backward state and resolved to do something about it. He evaluated recordings using a horn with a felt ear piece at the small end. At times he even bit into the wooden cabinet of the phonograph in order to "hear" more acutely the music being played. Because of his deafness and these techniques, he was able to hear overtones and wrong notes not readily recognized even by the artists involved. However, Edison's efforts apparently had little effect on the trend of music in the United States. He considered jazz only for "the nuts" and so concentrated on marches, melodies, and heart songs. As to opera recordings, he insisted that there be no tremolo—*i.e.*, quivering or trembling of the voice. In the case of one soprano, Edison suggested to his studio director that her breasts be taped down in an attempt to decrease her excessive tremolo.

During the 1908–1912 period the Edison Laboratories were intensively involved in research and development on all phases of the phonograph. Edison received 19 phonographic patents during 1909–1912, and many were also obtained by his associates. The phonograph was one field in which Edison encouraged his employees and close associates to work on their own ideas and obtain patents. It is likely that during this period he was so busy on his storage battery (Chapter 7) and the motion picture that he did not have time to formulate research programs in detail on the many phases of the phonograph which sadly needed attention. Edison organized the New Jersey Patent Company as a means of pooling patents of others and arranging for payment of royalties.

When Edison decided to introduce a disc phonograph, he sought the hardest surface material which could be impressed properly with a disc negative. He considered that celluloid was not hard enough and instead decided on phenolic resin, the first synthetic thermosetting resin which was at that time making its debut on the world scene. Phenolic resins are condensation products of phenol and formaldehyde. Although it was well known before the 20th century that such a chemical reaction led to solid resinous "gunks" with water as a by-product, it was 1905 before Leo H. Baekeland demonstrated that the phenol–formaldehyde reaction could be controlled to give useful products. Baekeland found that such resins were excellent binding agents for wood flour and that such compounds could be molded under heat and pressure to give strong heat-resistant parts. This discovery sparked the beginning of the phenolic resin industry which today produces over a billion pounds of product annually.

During this period Edison employed several chemists, of whom Jonas W. Aylsworth appears to have contributed most. Aylsworth was closely associated with Edison for about 30 years. Although in 1903 he retired as a full-time employee to become an independent consultant, he still continued to work largely with Edison. Beginning in 1909, Aylsworth made a series of improvements in the phenolics over those of Baekeland.* The most significant was the use of hexamethylenetetramine (often called hexa) for converting an A-stage phenolic—*i.e.,* one still soluble in simple organic solvents and fusible—to the B or C stage. Hexa is the condensation product of formaldehyde and ammonia. When used in proportion of 8–12 parts with 100 of fusible phenolic, the hexa converts the phenol on heating to a higher crosslinked stage. Such a blend was applied as a varnish to the base material of the record blank.

The original base for the discs was an asphaltic compound containing a high proportion of wood flour and china clay. Use of fillers served largely to eliminate shrinkage and thus warpage on cooling. Since the records were not used for successive recordings, the two phenolic surface layers could be quite thin. Adhesion problems were encountered in sealing the phenolic resin overlay to the asphaltic base, and the surface layer peeled after a time. This happens because asphalt is largely hydrocarbon in nature, and phenolic resins are highly polar; thus they are mutually incompatible. Furthermore, in the condensation of the soluble form of the resin to an insoluble state, any by-product water would tend to form a film at the interface of the two materials. The principal virtue in the use of hexa *vs.* paraformaldehyde (a solid polymer of gaseous formaldehyde) is that the formation of water and other gaseous by-products during the subsequent condensation is largely eliminated. However, it is actually impossible to eliminate

---

* Aylsworth's general contributions to Edison's chemical operations, and to the phenolic resin industry in particular, are discussed in Chapter 8.

completely the formation of such by-products in any thermal setting of a phenolic resin.

The adhesion problem was solved when wood flour-filled phenolic resin was used to make the record base. The proportion of hexa in the molding compound, plus the time and temperature used when molding, resulted in the phenolic binder's being converted to the C stage—*i.e.,* infusible and not softening with heat. The "Formula Book" (*12*) for record blanks shows that in 1925 Edison was still using a phenolic varnish to prepare the surface layer. Filled molding compounds for the base contained wood flour, ground chalk, lampblack for coloring, and either phenolic resin or rosin as binder.

Edison took out patents on several other features of thermosetting resins. One of these covered the use of phenylenediamine as a catalyst for increasing the crosslinking of shellac when heated above about 200°F, to convert it from a thermoplastic to a thermoset material. Phenylenediamine was also found effective for promoting the condensation of phenolic resin with hexamethylenetetramine. Edison's celluloid and phenolic recording surfaces were actually comparable with present-day vinyl records in providing maximum resistance to wear by the stylus with minimum tendency to plastic deformation. Since about 1945 almost the only material used for phonograph records has been a copolymer of 85–95% vinyl chloride and 15–5% vinyl acetate. Vinyl chloride imparts extreme hardness to the copolymer; vinyl acetate, flexibility. A wide range of hardness can be obtained by controlling the monomer ratio in the copolymer. The copolymer for records contains only 0.5–2.5% of a heat stabilizer such as an organic compound of lead, tin, barium, or cadmium and 0.2–2% of carbon black. The black pigment is used primarily to permit visual inspection. Translucent or transparent records are extremely difficult to inspect for imperfections.

Edison spared no expense in conducting studies related to

high quality recording. To record group renditions, he constructed huge horns of various shapes and sizes. One was 128 feet long, several feet in diameter at the large end, and weighed over a ton. It was made of plaster of paris with a very thick wall in the middle section. Another, made of brass, was 200 feet long. Thus even in the pre-microphone era Edison was able to record orchestras of up to 35 members and have each instrument register individually.

Several extensive historical collections of phonographs and records exist that represent the industry which originated largely from "Yankee ingenuity." That at the Edison National Historic Site is outstanding for Edison records. The inventory of "Diamond Discs" constitutes over 90% of those issued and includes about 2,000 recordings which have never been released. The collection of phonographs at the Museum of Thomas A. Edison's Winter Home, Fort Myers, Florida, is probably the most extensive in the world. The approximately 170 machines are all in working order. The collection includes the second tin-foil phonograph made by Edison. Robert C. Halgrim, Curator, continues to add to this outstanding collection. The Henry Ford Museum, Dearborn, Michigan, has an extensive phonograph collection, but because of limited exhibit space many of the present holdings are in storage. The phonograph collection at the Edison Historic Site is relatively meager, but it does contain the original tin-foil machine which was loaned to the South Kensington Science Museum in England in 1880. On request of the Edison Pioneers it was returned to the West Orange Laboratories in 1928 *via* the British Embassy in Washington. It is now on exhibit at the Edison National Historic Site.

Many museums and libraries have established archives to preserve old and new records for their historical value. Syracuse University has not only established an archive of recorded sound but is developing techniques to restore and re-record audio material. For a number of years, the Thomas Alva Edison Foundation and the Charles and Rosanna Batchelor Memorial Foundation have supported that university's Professor Walter L. Welch in his research dealing with optimum methods of re-recording Edison cylinders and discs. In the following quotation Professor Welch indicates why many of the early recordings and recording techniques are being reviewed (13):

We believe a return to some of the early idealism would be most wholesome. We have remarkably highly developed methods of recording and processing today, but we have lost sight of the early goals of precisely recording a performance exactly as it occurs and without attempting to improve it.

Although Edison no doubt made more technical improvements to the original phonograph than any other manufacturer, he did make several serious mistakes. Understandably, he was a very busy man during the decade following the original invention in 1877, but he should have found a way to continue some work on such an unusual and potentially important invention. His early emphasis on the use of the phonograph for business rather than entertainment allowed others to gain dominant positions in the more lucrative musical field. His continued preference for cylinders for records rather than the simpler discs was based on unrealistic quality considerations. Surprisingly, he failed to recognize the advantages of electronic recording and reproduction over purely mechanical methods, and this led in part to the demise of his phonograph business in 1929. It is likely that if radio had come in when Edison was 10 or more years younger, he would have been more receptive to this new technology.

Edison's development and manufacture of the phonograph made his name a household word even in rural communities. The autograph trademark "Thomas A. Edison" was inscribed on his phonographs and Diamond Disc record labels starting about 1900. His picture also appeared on some phonographs and record labels. During his summer camping trips with Ford, Firestone, and Burroughs beginning in 1918, people gathering at the stops would recognize him, if not the others. Should Edison ask the children if they knew who he was, he would usually get the reply, "Mr. Phonograph." One year at a stop near Bolar, Virginia, Edison lay down to take a nap in the shade of a tree. He awoke to find himself surrounded by a group of little girls. It must have been a rude awakening since one of them addressed him as "Mr. Graphophone."

Edison's work on materials for phonograph records helped usher in the "age of plastics" both from the standpoint of materials and fabrication methods. Although the soap–fatty acid compositions which he developed were used for making records for 25 years, they find no use in present day plastics. As in the case of most of Edison's chemical developments, nothing was published on these wax-like materials except in the patent literature. This class of materials may have been overlooked by chemists over the years as potential structural materials where high strength is not required. Edison played a major part in the early development of phenolic resins both in molding compounds and varnishes. However, the consolidation of this industry by acquisition in 1922 by Baekeland tended to overshadow the early contributions of Edison and others. Edison's use of phenolic resin as an overlay on disc records was one of the very few cases in which this thermoset plastic has ever been successfully used as a casting—*i.e.,* as 100% resin containing no fillers.

While working on this chapter in September 1970, the author noted in a local newspaper that McGraw-Edison Company had set up a new division known as the Edison Voicewriter Division. Previously, manufacture and sales of the Voicewriter had been a part of another division.

The invention of the recording and reproduction of sound is certainly one of the great scientific accomplishments of all time. This was emphasized by two of the speakers at the memorial program at the 1931 winter meeting of the American Association for the Advancement of Science honoring the memory of Thomas Alva Edison. Dr. F. B. Jewett, then Vice-President of the American Telephone and Telegraph Company, described this Edison achievement as follows (14):

In his invention of the phonograph Edison displayed an imagination, a skill and a perseverance of the very highest order. This invention alone might well have inscribed his name indestructibly in the history of America and of the world. At a time when phonographs, both acoustical and electrical, are an everyday commonplace, and where most of those in the world who are less than forty years of age can hardly conceive of life without the phonograph, it is difficult to appreciate the degree of daring which Edison displayed in even imagining that he could imprison such a fleeting thing as the energy of the spoken word. Equally difficult is it for the average man of today to conceive of the daring involved in imagining that out of the prison thus created could come, at some remotely distant time, a reproduction of ancient words, possibly those of men long dead, with the full vigor and clarity of the original speech.

Dr. Robert A. Millikan, Laboratory Director at California Institute of Technology and subsequent Nobel Prize laureate in physics, described the achievement even more dramatically (15):

The second scientific accomplishment is enough by itself to make any man immortal, for to Edison alone belongs the credit of conceiving and showing how mortal man may speak with his living voice directly to all the generations that follow after him. Could we today but hear Socrates and Marcus Aurelius and Shakespeare and Newton and Franklin and Goethe and Faraday and Maxwell, as our children and our children's children through the long ages will be able to hear their counterparts of today and of all the times yet to come, would we not build another Promethean legend around that deed akin to that of stealing fire from heaven and bringing it down to men. That man has lived and worked and walked on earth with us in our generation. Thomas A. Edison is his name.

IN THE EARLY 1880's the eastern iron ore industry found itself in a competitive situation that might be termed disastrous. The northeastern area of the United States had been an important source of iron ore during Colonial times, and blast furnaces were still being supplied from isolated sources of high grade ores in New York, New Jersey, and Pennsylvania, as well as from imported ore. The discovery and opening up of high grade iron ore mines in the upper peninsula of Michigan enhanced the competitive position of the midwest mills.

The entire Appalachian range contains iron deposits in various concentrations. Birmingham, Alabama, became a steel center, using local deposits of hematite, which is ferric oxide, Fe_2O_3. The major steel industry which developed there now uses primarily imported ore. The northern part of the Appalachians contains some hematite, but primarily the magnetite ore, Fe_3O_4. This iron oxide is unique in that its relative magnetic attractability is about 15, compared with 100 for iron; that for ferric oxide is only 0.77. The black tri-iron tetroxide is described as ferrosoferric oxide, $FeO \cdot Fe_2O_3$. However, its crystalline x-ray pattern and magnetic properties indicate a salt-like structure. The higher iron content of the Fe_3O_4, 72.36% iron compared with 69.9% for the Fe_2O_3, would appear to make it the more attractive ore. However, because of its greater thermal and chemical stability, it is more difficult to reduce to metallic iron than is Fe_2O_3; for example, in a present-day process to make spongy iron pellets, the Fe_3O_4 ore is first oxidized with air to Fe_2O_3 before it is subsequently reduced with hydrogen and carbon monoxide.

During the 1880–1885 period Edison apparently decided that the concentration of magnetite ore with magnets had become a potentially attractive field for research and development. His improvements on electromagnets for dynamos and electric motors no doubt made him feel that he had skills in this field that others did not possess. Also, he foresaw the advantages of the electric motor over steam-driven power in many of the operations in such a concentration process. Let us examine the technology of making iron and steel at the time Edison's interest was turning from electricity and the science of sound reproduction to that of supplying ore to this important industry.

Because of its chemical affinity for oxygen, iron occurs primarily as iron oxides in the earth's crust. Meteorites do contain metallic iron and probably were the source of the first iron used by man. To prepare iron from an iron ore—*i.e.*, from Fe_2O_3 or Fe_3O_4—it is necessary to reduce the oxide chemically. It has been known for centuries that iron ore, embedded in burning charcoal, can be reduced to metal. The ancient Assyrians are credited with the first recorded use of iron on an appreciable scale in about 3700 B.C. Their military advantage in iron weapons made them one of the most powerful nations of the then-known world. Shaft furnaces were developed to replace the crude hearths, and by 1350 A.D. they were widely used in Europe to produce molten high carbon iron which was molded into castings by pouring the liquid metal into molds of desired shapes. Because a series of small castings made in sand molds resembled a sow nursing a litter of pigs such iron became known as pig iron.

The blast furnace is essentially a large chemical reactor. It is operated substantially as it was centuries ago, although over the years technological improvements have made it a very efficient countercurrent fixed-bed process. Charcoal disappeared from use as wood supplies diminished, and it was replaced first in part by anthracite coal and later by coke. Coke is prepared by heating bituminous coal in a retort and must be physically strong to support the burden above it in the blast furnace. It is burned at the bottom of the furnace with a limited supply of hot air to produce primarily carbon monoxide:

$$2\,C + O_2 \rightarrow 2\,CO \text{ (Exothermic)}$$

Any moisture in the air reacts to form hydrogen and the monoxide:

$$C + H_2O \rightarrow CO + H_2 \text{ (Endothermic)}$$

The first of these reactions, being highly exothermic, furnishes the heat necessary to preheat the ore and melt the reduced iron. There are many reactions taking place in the body of the furnace, but the net result is that the reducing gases meet the iron ore at about 1700°F, the iron oxides are reduced to iron, the carbon monoxide goes to carbon dioxide, and the hydrogen goes to water. The molten iron is drawn off at the bottom of the furnace, and the impurities of the ore collect as a slag. In early blast-furnace operations, it had been found that use of limestone mixed with the iron ore served to form a liquid slag which also carries along with it most of the sulfur present in the ore. The limestone and other slag ingredients, being of much lower density than the iron, collect as an upper layer at the base of the furnace and are drawn off separately.

It was early recognized that the conditions under which iron was produced and subsequently treated affected its properties. The ancients developed what one might call a crucible steel by melting iron in a crucible and heating it to reduce carbon content or adding finely divided charcoal to increase the carbon content. The Bessemer process, invented in 1856, provided the first large scale method to refine pig iron to steel rapidly and cheaply under controlled conditions. In this process air is blown through molten pig iron in a vessel lined with a siliceous refractory. Oxidation of the silicon, manganese, carbon, and similar impurities is selective, and oxidation of the iron is avoided by controlling the amount of air. Also, proper control leaves the proper amount of carbon in the steel, usually about 2%, to give the desired properties.

The success of the Bessemer process depended upon the quality of pig iron available. This process did not remove either phosphorus or sulfur. Since both ingredients are highly detrimental in steel, specifications of pig iron for use in the Bessemer included a maximum of 0.090% for phosphorus and 0.030% for sulfur. This meant that the ore used for making the pig iron, as well as the limestone and coke, had to be low in these elements. The good reputation of Swedish steels down through the years has been largely the result of the low phosphorus contents of Swedish ores.

Others had attempted to concentrate various iron ores by jigging on screens in flowing water, thus separating impurities which tended to be more finely divided. The characteristic black color of magnetite made it possible to separate it manually from the light-colored rock impurities. In most magnetite mines, hand separation was economical if a man and shovel could be had for not more than $1.75 per day.

Magnets had long been used to separate iron from brass and other scrap from machine shop operations as well as nails, wire, and the like, from grains and cotton. Long before Edison began his operations at the Ogden Mine, manufacturers of the various magnetic separators and many hopeful inventors turned their attention to developing machines and processes for concentrating magnetite ores. As early as 1860 magnetite had been separated experimentally from sands along the St. Lawrence and on Long Island Sound. Such experimental work had involved the use of both electro- and permanent magnets.

At a meeting of the American Institute of Mining Engineers in February 1889, John Birkinbine presented a paper (1) co-authored by Edison, which stated that at least 10 different types of magnetic ore separations had been investigated in the United States, and that five of these had been operated on a semi-commercial scale. A Buchanan separator that they cited consisted of two iron rolls revolving towards each other on journals that were poles of an electromagnet; this magnetized the facing sectors of the rolls but not the opposite sides. Thus, finely divided ore passing through the rolls would be separated into non-iron tailings falling directly below the point of contact of the rolls while the magnetite would be carried around each roll until it passed over the non-magnetic portion; then it would drop off into separate containers. This equipment had been used with sea sand containing magnetite both in the United States and New Zealand, as well as with lean magnetite ore at the Croton Mine near Brewster, New York. The Wenström separator, of Swedish origin, consisted of a drum which turned around a magnet with ground ore fed on top of the rotating drum. Tailings fell downward while the magnetite concentrate was carried around about 180° and dropped into another container. Other magnetic separators were described using various belting arrangements.

Birkinbine's report also described in some detail what Edison had done prior to 1889 in this field. Edison's first patent dealing with the magnetic separation of substances by passing a falling stream of material past a magnet, was filed April 7, 1880 and issued 55 days later (2). The process was specifically developed for magnetite ore and involved crushing and sizing the ore, and then placing it in a hopper having a discharge opening which allowed a thin stream of material of considerable width to flow before the face of a magnet. The magnetite would be pulled away from the vertical stream and thus fall into a separate container. An electromagnet designed for this process was at that time on exhibit at the Edison West Orange Laboratories. The core was soft iron 6 feet long, 30 inches wide, and 10 inches thick, weighing 3,400 lbs. It was wound with 450 lbs of copper wire, coupled to a dynamo supplying 116 volts. The electromagnet consumed 16 amperes. Obviously, this massive magnet was not built until after considerable experimentation with smaller units.

Tests on lean ore ground and screened to 10 mesh, using a feed hopper with a 6-foot slot, indicated that 150 tons of material could be separated daily using one face of the magnet and 300 tons if both faces were used. Ores from eight different sources using the experimental magnet had been investigated at the West Orange Laboratory.

This paper also reported on a concentration unit which had recently been set up at Humboldt, Michigan, using lean magnetite ore from a dump-pile of 40–45% iron content. The ore, first crushed in a Gates rockbreaker, then "passes into a pair of 16 x 30 Cornish rolls, and from these the crushed ore is sized in screens to 20 mesh, the dust is blown out of it, and the product from these screens passes an Edison magnetic separator." The concentrate contained 62–68% iron and was within specifications on phosphorus content. It is not clear who financed this operation, but it was continuing in February 1889 at the time the paper was presented.

Edison had earlier tried to use his patented method for separating magnetite from sea sand *(2)*. In the 1880–81 period, despite the fact that all his available personnel were mainly active in developing the incandescent light, a company had been organized in an attempt to commercialize the ore concentration process. It was known as the Edison Ore Milling Company, Limited, and it was administered by the Edison Electric Light Company, S. B. Eaton, General Manager. Based on a series of handwritten letters by W. H. Meadowcroft (copies of which are at the West Orange Edison National Historic Site), it is also evident that an attempt was made to license the process. The following letter was sent to R. J. Edgar, 223 West 19 Street, New York City, on December 19, 1881:

In reply to your letter of 13th. . . . we will supply you and your associates the Edison Magnetic Ore Separator and Dynamo in such numbers as you may want, to be used exclusively on the property at Moisic, Canada. . . . Our prices and royalty for, say 10 separators, are as follows:

10 Magnetic Ore Separators	$10,000
2 Dynamo Electric Machines of sufficient power to run such separators	2,400
Wiring, etc.	500
	$12,900

Royalty, 25¢/ton of iron separated

Yours truly,

S. B. Eaton, Gen. Mgr.

Edison at that time had not carried the process beyond the separation step. The Edison Ore Milling Company set up a concentration unit utilizing sea sand at Quonocontaug Beach, Rhode Island, located a few miles east of Westerly, Rhode Island. Mr. M. R. Conley was in charge. A Meadowcroft message to him dated December 23, 1881 is quoted:

Mr. M. R. Conley, Quonocontaug, R. I.

Your letter of the 19th enclosing statement and vouchers. . . . duly received. I send you herewith $300, as requested, to pay bills and wages. Of course, you will keep your expenses as low as possible.

The following letter of January 10, 1882, indicated the operation was at least semicommercial:

To Wm. F. Block, Esq. Pres.

In reply to yours of the 2nd asking our prices in magnetic iron ore, . . . we will deliver same in cargoes of 50 or 100 tons at N.Y. or Jersey City, for $10/ton. We would supply you at this price continuously until we gave you 30 days notice to

144

the contrary. As output is at present limited in amount, we could not now supply you more than 150 to 200 tons/month, but we expect to be able nearer Spring to supply a much larger quantity. Our ore has been used by the Poughkeepsie Iron and Steel Co. and we are informed that the iron made from it has been sold on the market and reported upon favorably. . . .

You ask if we would entertain a proposition for your Company to work an ore bed on royalty. If you have such an ore bed, we would certainly be willing to supply our machines for you to work the same.

Apparently, however, there was not too much progress, based on a letter of March 10, 1882, Eaton to Conley:

Please do not spend any more money than is absolutely necessary at present. We are receiving no income yet and I am afraid our Directors will grumble if our expenses are too heavy. . . .

Other beaches were searched unsuccessfully to find sand of higher magnetite content. The Poughkeepsie company apparently was the only customer, and they stopped purchasing the concentrate during the latter part of March 1882. Although Conley and a Mr. Beebe continued to work at the Rhode Island site through 1882, early in 1883 the buildings were sold, and the separator was packed and shipped to the Edison Machine Works, Goerck Street, New York City. In a letter to R. L. Cutting, Jr., Treasurer of the Edison Ore Milling Company, Eaton on January 15, 1883, advised: "Our ore separating business at Quonocontaug, R.I. has been closed and according to the records $2,400 is due." Presumably Edison owed that amount to the Edison Electric Light Company, thus concluding his first venture in iron ore concentration at a loss.

Birkinbine again addressed the American Institute of Mining Engineers in September 1890 on the subject of magnetic separation of iron ore (3). At that time he was president of the Institute, and one can assume he was quite knowledgeable of his subject. Birkinbine attempted to answer the question as to whether such a concentration process was economical. He cited the experiences with a number of magnetic separators, including that of Edison's. He estimated that in 1890 approximately 90,000 tons of concentrated ore would be marketed, about half of which would be separated by magnetic methods and the other half by wet jigging. Birkinbine felt that loss of iron in the tailings from magnetic separation might be a serious problem. In one installation using a Conkling magnetic separator involving a wet process using an endless belt, the following data were cited:

Feed	26.8% Fe
Concentrate	50.2
Tailings	11.5

He estimated cost per ton of concentrate at $2.25, with only 61% of the iron recovered. In the New Jersey–New York area it appeared more economical to ship ores as low as 38% iron and use more flux in the blast-furnace operation than to concentrate them. In the Michigan area, where freight charges were higher to transport the ore to market, ores as high as 52–56% were being concentrated.

The magnetically separated ores were being supplied as fines and were mixed with mine-run material when charged to the furnace. Birkinbine concluded in his paper at that September 1890 meeting that although magnetic separation of iron ores was well beyond the experimental stage, the volume of production was as yet too small to attract the general attention of consumers.

Interest in magnetic separation was also emphasized in several other papers presented at the September 1890 meeting of the Institute. In his presentation entitled "The Ball-Norton Electro-Magnetic Separator," C. M. Ball, stated in his introduction that (4):

The magnetic concentration of iron ores has been so often and so widely studied and discussed among the members of the

Institute that any remarks concerning its general importance, from an economic standpoint, would be superfluous.

Edison's failure with sea sand as a source of magnetite did not change his conviction that the approach using ore was feasible. This would involve mining, crushing, grinding, and screening, as well as separation. By 1885 he had found time to plan his operations using ore. As was the practice of others during that period, Edison set up his separator and auxiliary equipment at magnetite mines to process the ore which had been mined but which was too lean in iron content to market economically. Conventional crushing, grinding, drying, and screening equipment was used, and the concentrate was shipped as fines. In his review in September 1890, Birkinbine cited five concentrating units operating in New York, one an "Edison machine" at West Point (*3*). Presumably this was a licensed unit. Relative to the previously mentioned Edison operation at Humboldt, Michigan, the gangue proved to be of such a nature that the operation was both difficult and expensive. When the plant was destroyed by fire, it was not rebuilt. This was early in 1890. However, this plant demonstrated that the Edison separator was capable of producing a very high purity concentrate. A carload of concentrate analyzing 71.4% iron was prepared there. One hundred percent pure Fe_3O_4 contains 72.36% iron.

We also know that in the late 1880's Edison set up a concentration unit at Bechtelsville, Pennsylvania, about 15 miles northeast of Reading. Here he used low grade virgin ore, and the plant consisted of crushers, rolls for grinding, screens, and Edison separators. A total of 1,000 tons of concentrate was delivered to local blast furnaces. By 1890 the operation had been shut down since "the uncertain character of the deposit did not warrant the continuance of operations" (*3*).

By 1889 Edison concluded that magnetite as low as 20% iron content could be economically concentrated if operations were on a large scale; this was based on his laboratory data and small plant operations. Thus, as in the subdivision of electricity for lighting, we find him ignoring opinions of the experts that magnetic concentration of iron ores, even of considerably higher iron content, was of doubtful feasibility. No blast furnace had ever been run on concentrates alone, and most furnace operators didn't like such ores if they were finer than about 5 mesh.

With the formation of Edison General Electric in 1889 he had received a large sum of money. To him the purpose of money was primarily for research, development, and venture capital, and magnetic separation of ores was high on his priority list.

In 1889 there were 14,500,000 long tons of iron ore produced in the United States. This total was distributed as to types as follows:

Magnetite	17.5%
Brown hematite	17.5
Red hematite	62
Carbonate	3

Production of the magnetite was distributed geographically as follows:

New York	37%
Pennsylvania	34
New Jersey	16.5
Michigan	10
Others	2.5

New Jersey's production of total iron ore in 1889 was less than half that of 1882.

Although magnetite production was low, and that of New Jersey represented only 2.9% of the total U.S. ore produced, Edison realized the potential was much greater if low concentration ores could be utilized. He had a number of men on

148

foot make a geological survey with magnetic dip needles in search of magnetite deposits. Starting at the St. Lawrence and going to the Potomac, the Allegheny range was criss-crossed at distances of approximately one mile. When a body of magnetite was found, it was checked in a parallel manner at intervals of 100 feet. Several large deposits were found, including one near Ogdensburg, New Jersey. By purchase and lease, mineral rights of 16,000 acres were acquired. This became the site of Edison's fabulous demonstration plant.

A company known as the New Jersey and Pennsylvania Concentrating Company was organized, and financing was arranged. As it turned out, the bankers proved skeptical of the project, and Edison himself largely financed it. Ogdensburg is situated only about 40 miles from the northern tip of New Jersey and is in a beautiful area of highlands and lakes. The plant was built on a ridge about two miles east of Ogdensburg and the village surrounding the plant became known as Edison. A few stores were opened, followed by a post-office and a school. A town named Summerville sprang up around Edison where many of the miners lived. At the height of Edison's operations, the two towns were said to have had a population of about 1,000.

Edison had estimated that there were 200,000,000 tons of low grade magnetite ore in the immediate 3,000 acres. This included the Ogden Mine which had been worked off and on for about 50 years. Early Dutch settlers had found iron, copper, and zinc ores in the adjacent valley. The New Jersey Zinc Company was located at Franklin, New Jersey, only about four miles from Ogdensburg. Fortunately, a railroad ran along the base of the mountain where Edison was located, and a spur was built to his plant.

The Ogden Mine had been operated as a surface mine with several narrow pockets of ore 3–12 feet wide and 200–300 feet long. Operations had reached depths from which it was practical to pump water. Edison's operations were to be in the

149

same area, but using a strip 300 feet wide, with no attempt to segregate the ore prior to concentration. He said he would process a "whole mountain." His initial approach apparently was to employ massive handling equipment to transport the ore cheaply to the crushers. Dynamite would be used at a minimum since "energy in coal at $3/ton is a lot cheaper than that of 45% dynamite at $260/ton." Originally he planned to use conventional equipment from the crushers on through to the separation step as he had at his previous installations, and to sell the concentrates as fines. In his report of September 1890, Birkinbine stated that the construction of Edison's plant at the Ogden Mine was nearly completed (3). In a series of papers presented before the Institute at its Cleveland Meeting in June 1891, several of the speakers referred to Edison's operations at Ogdensburg. Birkinbine (5) commented that Edison was grinding and screening to 60 mesh, and "based on good authority" Birkinbine understood 100,000 tons of this material would be sold for blast-furnace operations. A representative of the Lackawanna Iron and Coal Company (6) stated they had been using magnetite concentrates for six years in their furnace operations. With $1/4$-inch particles they had been able to use such a concentrate for up to 50% of total ore fed. Lackawanna Iron and Coal Company had used 117 tons of 50-mesh concentrate from Edison, and this had tended to fly when unloaded from cars dumped into the furnace. This was overcome by moistening the ore before unloading. This particular ore apparently came from the Ogden works since Edison had not gone to such fine material in his previous operations.

It is clear that the Ogden plant was designed to function both as a pilot plant and ultimately as a truly commercial plant. As a pilot plant the objective was to demonstrate that magnetite ore containing about 20% iron could be concentrated to an acceptable feed to produce pig iron in a Bessemer blast furnace and at a price competitive with high purity mine-run ores of a comparable iron content. As is the purpose of any large pilot plant, sufficient material would be prepared to allow large scale evaluation of the product by potential customers. In this case it meant enough product to allow blast-furnace runs up to 100% of concentrate.

Edison was apparently so sure his project would be a commercial success that many features of the plant were built much larger than necessary for a true pilot-plant operation. This was a basic mistake. Accepted practice today in developing new products and processes is never to build a pilot-plant unit larger than necessary to demonstrate the process and to prepare adequate product of commercial quality for evaluation by potential customers.

It is not within the scope of this chapter to discuss Edison's iron ore plant in detail. However, since it had so many ingenious features compared with practices at that time, we cannot fully appreciate the chemical and chemical engineering contributions which he made here without getting a general overall picture of the entire operation.

The man who a few years before was working with fragile carbon filaments so delicate they required the skilled touch of a watchmaker now turned to a project where massive machinery expertise was the guideline. The ore-concentrating plant at Edison was designed so that the ore need not be touched by hand from the quarry to the freight car for shipment. The plant was set up in three basic units, each of which could operate independently of the other two. These operations were as follows:

(a) Quarrying and crushing the ore to ½-inch or less in size, drying, and then storing in a stockhouse of 16,000 tons capacity.

(b) From this storage, the ore went to grinders and then to magnetic separators. The concentrate went to a stockhouse and the tailings to an outside dump.

A view of Edison's ore-concentrating plant demonstrating its large size

A view from the other side of the plant. The large power plant in the right foreground further illustrates the magnitude of the operation.

(c) A scraper conveyor carried the concentrate to the briquetting operation. The briquettes were baked and either stored or shipped.

A lumber mill and various shops were set up as the base for constructing the plant buildings and repairing the equipment. Most of the custom-made equipment was fabricated at the laboratory shops at West Orange. As an example of what could be made, a traveling crane was built which was capable of handling 10 tons. Fifty houses were built at Edison for the families of skilled mechanics and other key personnel. This was the era when workmen had to live within a few miles of their job. Each house had running water and electricity although the neighboring villages were still burning kerosene and candles. The workmen's wives liked their modern conveniences and the fact that their husbands spent more time at home than previously. There wasn't much to do in this rural area; one was either at home or at work.

Edison brought in no mining experts but proceeded on his own to design and build his plant. His approach to mining lean iron ore and concentrating it to acceptable standards was in direct contrast to the practice at that time of mining only concentrated deposits. Editors of *Iron Age* magazine visited his plant in 1897. The following quotation dealing with their visit is taken from the publication (7):

But to our mind originality of the highest type as a constructor and designer appears in the bold way in which he (Edison) sweeps aside accepted practice in this particular field and attains results not hitherto approached. He pursues methods in ore dressing at which those who are trained may well stand aghast.

Phase 1 of the plant operation called for handling 6,000 tons daily. In line with using a minimum amount of dynamite, a series of 3-inch holes, 20 feet deep, were drilled 8 feet apart and 12 feet back from the ledge. When these were filled with dynamite and the charges set off simultaneously, up to 35,000 tons of rock were dislodged. A 93-ton steam-driven shovel, said to be the largest in America and formerly used in dredging the Chicago drainage canal, loaded the rock onto skip cars. These ran on a narrow gauge railroad by gravity to an area where a traveling crane moved them to the giant crushing rolls (*8*). This machine had a total weight of 130 tons, and the two solid cast-iron rolls, 6 feet in diameter and 5 feet long, were covered with chilled iron plates containing a number of projecting knobs 2 inches long. The surfaces of the rolls were 14 inches apart, and the total moving parts weighed 70 tons. The rolls turned toward the opening as in case of a typical mill used for masticating rubber. The rolls were belt driven through friction clutches and would be rotating at a circumferential speed of nearly a mile a minute when a skipload of rock was dropped on them. The pile-driving effect of the rolls crumbled rocks weighing up to 6 tons. The power was released with a clutch at the moment of resistance, thus causing no strain on the drive. The crushing decreased the momentum of the rolls about 10%, but as soon as a given charge passed through the rolls, the belt drive would again come into play and in less than a minute speed up the rolls for the next charge. By using momentum for shock crushing, a much smaller steam engine could be used for power than with a slow rotating conventional crusher.

The rock, thus reduced to 14 inches or less in size, then passed through smaller crushers of similar construction but positively driven. After the rock was reduced to pieces ½-inch in size or less, it was elevated to the top of the dryer. This patented device was a tower 9 feet square and 50 feet high heated from below by an open-furnace fire (*9*). The tower contained a series of layers of cast iron plates, 9 feet long by 7 inches wide, arranged alternately at 45° angles in a "fish-ladder" fashion. The crushed rock would fall from plate to plate, meeting the hot air in a countercurrent manner. It fell from the dryer onto a continuous conveyor which took it to the stockhouse.

This method of drying was unique with Edison. The best dryer for coarse aggregate on the market at that time was the rotary type with a maximum capacity of about 20 tons per hour. Edison designed his tower dryer for 250 tons per hour and was able to demonstrate a capacity of 300 tons per hour under dry weather conditions. It was largely a matter of bringing the ore to a temperature sufficient to remove surface moisture. The dryer had no moving parts. The fines and shape of particles had to be controlled so the crushed ore would move down over the inclined plates by gravity alone. Occasionally the ore plugged the dryer. One time Edison and his plant superintendent, Walter S. Mallory, got into the tower through a manhole to examine the problem, and several tons of ore dislodged and embedded them. Fortunately, no injuries resulted. The Old Man had the reputation that he would never send a man anywhere he wouldn't go himself.

This type of dryer would appear to be very simple to construct and required no power except that to elevate the material to the top. It would be suitable only for coarse free-flowing material. This type of dryer is not generally used today. The present-day drying technique nearest to Edison's gravity countercurrent flow with hot air is probably the use of a fluid bed. Here the particles must be of a size comparable with that of sea sand, and they must behave substantially as a liquid when aerated at the base of the bed.

The drying and subsequent operations were carried out under well-housed conditions. The crushed ore was ground through three-roll mills with the rolls arranged one above the other. The rolls were made of cast iron and were 36 inches in diameter and 30 inches wide. Designed and built by Edison, they were said to utilize the power consumed much more efficiently than grinders on the market at that time. The lowest roll was set in fixed bearings while the other two were free to move up and down in guides with adjustable pressures from an air cylinder arrangement. The degree of fineness could be adjusted, and the contact pressures of the rolls varied depending on the hardness of the rock being ground and rate of feed. The ore was fed to the upper rolls and then passed to the lower ones. There were four such three-roll mills; two were usually used with virgin feed and two on recycle from the screens.

The ground material went to 14-mesh screens. Theoretically, such a sieve has openings of 0.0557 inch which means that the powder going through is a little smaller than 1/16 inch at its maximum cross-section. Edison pointed out in his basic patent (*10*) that previous use of stationary screens had been unsuccessful; when they were tilted enough to let the powder flow by gravity without plugging the openings, it went too fast for effective separation. What Edison did was to use short screens inclined 45° with baffles to arrest the velocity of the particles before they passed on to the next layer of screens. Also, the screens had slots in the direction of flow rather than square holes as do conventional screens today. Theoretically a sieve of such slot construction could let particles of oblong shape pass through which were longer than the desired 14 mesh. However, since the particles rolled down over the slots, the longer dimension would tend to be crosswise to the slots—*i.e.,* a short cylindrical particle tends to roll rather than tumble end over end.

Unlike the conventional rotary screens of that time, this stationary process used no power except to elevate the material to the top of the multiple screen equipment. When screening relatively fine material, it was necessary to mix the powder with a high proportion of coarse material to avoid plugging (*11*).

Although Edison investigated and patented many different techniques for applying magnetic attraction to concentrate magnetite ores, he considered only the falling-bed technique at his Ogden project. A total of 480 electromagnets were used in three series of operations employing so-called 12-, 8-, and 4-inch magnets. The 12-inch type was made with a cast iron core, 4½ feet long, 12 inches wide, and 4 inches thick. This was considerably smaller than the 30-inch magnet previously exhibited and demonstrated at his West Orange Laboratory. Each was wound with No. 4 copper wire. A direct current of 80 volts was used, and the windings of the various magnets were varied to regulate strengths. For a given falling stream, relatively weak magnets were first used to remove the high concentration particles, followed by stronger magnets to remove less pure particles which were reground, screened through 50-mesh screens—*i.e.,* to particles of 0.0117-inch maximum size—and then recycled through the separators. The concentrates were progressively increased in iron content in the three series of separations, and the tailings were either recycled with the lean ore or discarded once the iron content reached about 1%. The fine concentrate consisted of 67–68% iron—*i.e.,* 94–95% magnetite—indeed a very high purity iron ore. Also, the loss of iron in the tailings, which were sold to the construction industry, represented only about 4% of the iron originally mined.

Phosphorus in iron and steel imparts brittleness, and its concentration is rigidly controlled in all iron and steel products. In Edison's process of concentrating magnetite as a powder, it was possible to introduce a simple purification step to remove most of the phosphorus that had not been removed by magnetic separation. Because of the lower hardness of the phosphorus compounds, consisting largely of calcium and magnesium salts, these tended to be ground to finer particles as compared with the iron oxide. This fact, and because the magnetite is about 1.7 times as heavy as such calcium compounds, made it possible selectively to blow the phosphorus from the magnetite powder. This was done in a tower-type arrangement. The fines removed in the aeration process were recovered and sold for use in paints. The concentrated ore analyzed only about 0.03% phosphorus.

Once Edison determined that he had to grind and screen the ore to a fineness of 50 to 60 mesh to obtain a high iron content concentrate and low iron in the tailings, he realized he would have to agglomerate the concentrate before marketing. Only briquetting was considered for preparing the agglomerates (*12*), according to the following excerpt from *Iron Age* (*7*):

> While Mr. Edison and his associates were working on the problem of cheap concentration of iron ore, an added difficulty faced them in the preparation of the concentrates for the market. Furnacemen object to more than a very small proportion of fine ore in their mixtures, particularly when the ore is magnetic, not easily reduced. The problem to be solved was to market an agglomerated material so as to avoid the drawbacks of fine ore. The agglomerated product must be porous so as to afford access of the furnace-reducing gases to the ore. It must be hard enough to bear transportation, and to carry the furnace burden without crumbling to pieces. It must be water-proof, to a certain extent, because considerations connected with securing low rates of freight make it necessary to be able to ship the concentrates to market in open coal cars, exposed to snow and rain. In many respects the attainment of these somewhat conflicting ends was the most perplexing of the problems which confronted Mr. Edison. The agglomeration of the concentrates having been decided upon, two other considerations, not mentioned above, were of primary importance—first, to find a suitable cheap binding material; and, second, its nature must be such that very little would be necessary per ton of concentrates.

It proved to be a long and difficult task to develop a proper binder, learn the best way to mix it with the fines, and to fabricate the briquettes at low cost. A laboratory was set up at the concentrating works primarily to work on this project. On August 24, 1891 Edison filed two patents using combinations of lime and clay, or rosin (as sodium resinate) and clay, as binders (*13*). Although the rosin softened in the subsequent blast furnace, the clay held the particles in the briquettes together until they were reduced to iron.

These binders were added with just enough water to prepare a slurry which could be handled. Edison also tried mixing the rosin with the ore dry and then wetting with petroleum to make it gunky enough so that the mixture could be compressed. At one time he had a fire in the ovens used for baking the briquettes, so we can assume that he at least experimented with the petroleum softener. It is not clear what binder was standardized, but it was probably some organic material in combination with a bricking clay. Rosin appeared to have been of considerable interest, but after sending a man through the South to study the rosin supply from the naval stores industry, Edison decided that the supply would be inadequate for such a potentially large use. It has been rumored that molasses was used with clay. The molasses would decompose to carbonaceous materials in the baking step and help to make the briquettes water resistant. Bentonite clay in 1–1.5% concentration, plus a few percent of powdered coal, is now commonly used when making pellets of fine magnetite concentrate.

Edison mixed the fine concentrate in horizontal mixers with the binder added at intervals. A scraper-type conveyor carried the pasty material to the 30 briquetting machines. Each machine had a rotating die block which pressed out cylindrical cakes, 3 inches in diameter and 1½ inches thick, at the rate of 60 per minute. The briquettes were compressed under successive pressures of 7,800, 14,000, and 60,000 pounds. They then dropped into bucket conveyors and were carried through vertical ovens where they were baked at 400°–500°F for about one hour. Each conveyor made five loops through an oven, of which there were 15, one for every two briquetting machines.

The briquetting machines were built at Edison's shops in West Orange (14). Since 30 were used in the operations, this must have been a major construction project as well as one of intricate design.

A novel feature of the plant was the lubrication system. With dusty operations and many bearings operating under high loads, major wear problems would be expected. This was largely avoided by operating all stationary bearings and gears immersed in oil. Oil continuously flowed into each working mechanism, and then drained into a reservoir from which it was pumped to filters for reuse. Oil from a central supply of about 4,000 gallons was piped to the buildings in underground lines. Each building had an elevated tank for gravity feed to the machines.

Edison's handling of materials was probably the most automatic of any similar industrial operation up to that time. Henry Ford later said that reading articles on Edison's Ogden Mine project had given him the idea for his production-line technique for assembling Model-T Fords. Edison played a major part in the development of what we know today as the rubber belt conveyor. Actually, it should be called a rubber–fabric belt since the fabric is what gives the belt its strength, the rubber serving only to bind the fibers of the fabric or cord together and to protect them from abrasion and other destructive influences.

In the spring of 1891, a young salesman handling water hose and other rubber industrial products called at the Ogden Mine (*15*). He found Edison conveying crushed ore with canvas belts which ran on simple wood rollers rigged with side-boards along each side to keep the ore from falling off. The belts were coated with a waterproofing compound which also protected the canvas to some degree from abrasion by the ore. However, the belts lasted less than two months.

The salesman, 23-year-old Thomas Robins, suggested that rubber might be better for coating the canvas. At that time rubber was mainly used for waterproofing boots and raincoats. Little was known of its toughness. In cooperation with Robins, tests were set up at Edison's laboratory to determine the abrasion resistance of vulcanized rubber. They proved so encouraging that Edison gave Robins an order for 500 feet of canvas belting with a ⅛-inch thick rubber cover on top. The new belting was made by an outside manufacturer and proved quite successful. Robins and Edison became friends, and the young salesman-turned-inventor worked in Edison's laboratory at the mine and tested experimental products in the conveying systems. Rollers were developed on angle brackets in such a way as to raise the sides of the belt and form a trough for the ore to ride in. This is the basic design of the modern conveyor belt.

In June 1896 Robins formed the Robins Conveying Belt Company. This company made the auxiliary equipment and had the belts made to specification by the Hewitt Rubber Company. In 1945 the two companies merged to form Hewitt-Robins Incorporated, now a part of Litton Industries, a major supplier of complete belt conveyors.

Edison's installation on the 1,200-foot ridge east of Ogdensburg was well underway by 1892. The plant grew in a topsy-turvy fashion with changes being made in any operation that did not live up to expectations. Operation of the huge crushing mill had to be tamed down a bit since in early trials boulders as large as 1,000 lbs were thrown 25 feet into the air. Since the wide use of conveyors required that the buildings which housed the operations be specifically designed, a change in operations might make a building obsolete overnight. In one case, a building had just been finished, and even before any equipment was installed, Edison ordered it torn down, foundations included, and a new one erected on the same site.

A low volume of production had been achieved by 1893, and small shipments were being made to various blast-furnace operators when the panic of 1893 struck. Industrial operations all over the country halted, and Edison had difficulties in meeting his payroll. Instead of shutting down operations completely, he stopped producing concentrate and went through a series of developments, rebuilding and improving most of the operations. By late 1896, with the company again rapidly moving ahead, Edison was ready to try full-scale operating runs.

Early in 1897 approximately 500 tons of the briquettes were

shipped to the Crane Iron Works at Catasauqua, Pennsylvania, which is in the Bethlehem area, for an extensive blast-furnace test. The furnace in which the test was carried out produced 100–110 tons of pig iron per day. The magnetite briquettes made up 25% of the ore feed the first day, and this was increased incrementally until the briquettes made up 100%. At this point the yield of iron was 138½ tons for the 24-hour period tested. The percent briquettes was then decreased in steps and when the furnace was back on standard feed, the yield leveled out at regular production.

Although the test was not long enough to be wholly conclusive, the results were highly encouraging. The yield of the blast furnace was not only increased about 30%, but the limestone requirement was reduced from 30% to 12% of the ore charged and fuel consumption decreased. The phosphorus content of the resultant iron was reduced from the usual 0.9% to about 0.15%, thus resulting in a stronger and tougher metal.

In spite of the favorable results at the Crane Iron Works, Edison had difficulty in getting iron companies to purchase his ore concentrate. Let us examine what was happening in iron ore developments elsewhere while Edison was crushing and concentrating lean magnetite ores in New Jersey.

The Saint Marys River connects Lake Superior and Lake Huron and is the boundary between Ontario and Upper Michigan at that point. The rapids of Saint Marys make it unnavigable. On the United States side, a railroad was first used to carry passengers and freight around the rapids. In 1844 a U.S. Government survey located a large deposit of magnetite in Upper Michigan, just south of Lake Superior. Congress authorized the building of a canal between the two lakes, and this was undertaken by the state of Michigan. It was a tremendous job.

The nearest telegraph was at Detroit, 450 miles away. Blasting powder had to be brought in from Delaware. All drilling was by hand, and excavation was by scoops drawn by horses or mules. Much hand labor was required, but since little labor was available in this wilderness area, recent immigrants were brought in from New York and Boston. The labor force reached a maximum of 2,000. Winter temperatures fell to $-35\,^\circ$F, and disease took its toll. During the two-year period, 200 men were lost, mainly because of a cholera epidemic.

Iron ore was brought down to Marquette, Michigan, by mule team on plank roads. On August 14, 1855 the *Columbia* took on 100 tons of the rich black ore, and on August 17 this first shipment of Lake Superior ore passed through the canal. This canal had two stone locks, each 350 feet long and with a lift of 9 feet. Lake Superior ore then began to move in quantity to Ohio ports and thence to Pennsylvania and eastern markets. By 1857, 1,000 tons of ore daily were being delivered to the Marquette docks.

In 1881 the U.S. Government took over the Soo Canal (by the Saint Marys River). An improved lock was built soon afterwards, and in 1896 the highly improved Poe lock was completed at the site of the original set of locks. The Poe lock is 800 feet long and has 22 feet of water on the wooden sills against which the gates are closed. The canal was widened in 1908, and there are now four locks on the American side.

The Canadians were not idle. In 1888 they began a 1.3 mile canal running through Saint Mary's Island on the north side of the Saint Marys River. It has a single lock 900 feet long and 60 feet wide. The project was completed in 1895, and for many years the larger ships used the Canadian canal because it was wider and deeper than the American. Thus we see that in the 1894–96 period the Soo Canals were no longer a bottleneck to the passage of lake steamers, and bigger and bigger ships could be handled on an increasingly efficient basis.

The demand for steel during the Civil War consumed all the pig iron available. Mills sprang up in the Chicago area, and those in Ohio and Pennsylvania expanded. Blast furnaces at Pittsburgh and Johnstown used all the ore they could get. In 1866 a total of 240,000 tons of ore moved out of Marquette alone.

The ore from Upper Michigan was mostly magnetite, coming from shaft mines less than 200 feet deep although some went as deep as 500 feet. New discoveries were made at Menominee, just above Lake Michigan, and a railroad carried the ore to Escanaba, a Lake Michigan port. Ore discoveries spread over into Wisconsin in the Vermilion Range.

However, the Lake Superior area did not give up for long its position as the No. 1 producer of ore. In 1865 ore was discovered near Lake Vermilion, Minnesota, about 70 miles from Duluth. Shipment began, but development was slow. Minnesota was much more a wilderness during this period than Michigan or Wisconsin. Whereas Michigan became a territory in 1805 and was admitted into the Union in 1837, Minnesota was not established as a territory until 1849. Minnesota became a state in 1858, 10 years after Wisconsin was admitted. Ore production at Lake Vermilion was underground and was competitive with that produced in the Vermilion Range in Michigan and Wisconsin.

The lumber industry moved from Michigan into Minnesota during the post-Civil War years. The Merritt family, which included five sons, were leaders in the logging industry. While working on the Mesabi Range, they had noticed outcroppings of "soft red stuff" and took out claims. As the logging operations moved farther west, the Merritt brothers' interests turned to iron ore. Pooling their meager resources, they brought in an experienced mining man, Captain J. A. Nichols, to drill exploratory holes. On November 16, 1890, he struck the first large deposit of soft hematite ore in Minnesota. A 50-lb sample taken to Duluth analyzed 64% iron, which meant it was over 95% Fe_2O_3. However, mining "experts" advised that the find was no good since it was so loose one couldn't sink a shaft into it.

Leonidas Merritt, the oldest brother, saw no need for shaft mining and set up several home-made steam shovels to recover the loose ore. This was truly an open pit operation whereas Edison's operation would be called open-cut mining. Merritt was able to interest Henry Oliver, the Pittsburgh plow manufacturer, who in turn interested Henry Frick who was Andrew Carnegie's righthand man. The Merritt brothers borrowed heavily to build a railroad to Lake Superior, about 70 miles away. Work was well underway when the panic of 1893 came, and sales of ore and sources of credit ceased to exist. They were able to borrow $250,000 from John D. Rockefeller by giving the bonds of their railroad as collateral. More debts had to be met, and they appealed to Rockefeller for additional money. By 1895 the largest iron ore deposits in the world were in the hands of Rockefeller's Lake Superior Consolidated Iron Mines for a total expenditure of about $420,000.

Oliver had obtained a lease to take the ore from the Mesabi Range developed by the Merritt brothers, and this was shared with Frick and Carnegie. Rockefeller had an interest in the American Steel Barge Company, which moved ore on the Great Lakes. Irrespective of one's personal feelings for the unhappy fate of the Merritt brothers and other pioneers who brought in the Mesabi discovery, the stage was now set for rapid development of this fabulous deposit. Financing certainly was no longer a problem. The Standard Oil Company had created a reservoir of money, and Rockefeller was looking for new investments. Carnegie had the blast furnaces, the steel mills, his own coke, connecting railroads, and interests in several Great Lakes shipping lines. This situation, plus an open highway now available for vast shipping through the Soo

Canals, heralded a new era in iron ore production in America. By 1896, 20 open mines were operating on the Mesabi, producing nearly three million tons of ore a year.

Production costs on the Mesabi proved to be lower than anything Edison had dreamed possible. Since the usual magnetic needle could not be used with hematite ore, fortunately the diamond drill had come into use and thus facilitated exploration for this ore. The soil overtop proved to be quite thin and the ore deposits were very deep. Steam-driven shovels capable of taking 13 tons per bite were brought in. By 1898 the production of the Mesabi Range surpassed that of Upper Michigan. By 1900 there were 111 pits operating on the Mesabi with ore being delivered to the Duluth area ports at a selling price of 50¢ per ton.

By early 1897 it was apparent in mining circles that the price of iron ore would remain low for some time. Bessemer pig was about $11 per ton, compared with $17 a decade before. The lower price was primarily the result of cheaper and better ores. At the February 1897 Meeting of the American Institute of Mining Engineers, E. Gybbon of Trenton, New Jersey, presented a paper entitled "Improvements in Mining and Metallurgical Appliances during the Last Decade" (*16*). Included were some brief remarks about the "remarkable work in the mining, crushing, concentrating, and finally briquetting of magnetic iron ore which is carried on at the Ogden Mine in New Jersey by Mr. Thomas Edison." He went on to say that if ore used for concentrating contained 30% iron, the cost of iron in the final product would be about 6¢ per unit. He concluded his remarks about Edison's Ogden project as follows:

Until the price of foundry-pig shall advance to a figure at which it will be profitable to purchase a 68% iron ore at 6¢ per unit (i.e. $4.08 per ton), it is not probable that the Edison Works can be run continuously at a profit. It stands, nevertheless, a monument of perseverance in original research which certainly deserves our admiration.

At the July 1897 Meeting of the Institute, held in the Lake Superior region, John Birkinbine presented a paper on "Iron Ore Supply" (17). He stated that ⅝ of the total U.S. supply came from the Lake Superior region and that in spite of an 800-mile shipping distance, iron ore of 63% iron content and as low as 0.045% in phosphorus was being delivered at lower lake ports for only $2.40 per ton. To illustrate the rapid growth of ore production from the Mesabi Range, he cited the fact that the first shipment had been made in 1892, while in 1896, 3,082,973 tons of Mesabi ore had been marketed. There were many additional papers presented at this meeting on the Mesabi Range operations.

However, one must not forget that magnetite had been considered the preferred type of ore by East coast blast-furnace operators. Edison felt his concentrate would command a major premium price over the loose Mesabi hematite. In fact, first reports from the field were that the Mesabi ore behaved poorly in the Bessemer converter. One fact strongly in Edison's favor was that he could produce a concentrated ore of very low phosphorus content regardless of the amount of phosphorus in the lean ore. At that time, many ores of adequate iron content were unsuited for blast-furnace reduction because of high phosphorus contents. However, the Gods of Technology appeared to be against him since a new furnace began replacing the Bessemer in the 1890's, which was capable of producing low phosphorus steel from high phosphorus pig iron.

The Bessemer was the principal steelmaking process in the United States until 1908. Karl Siemens, a German-born naturalized British citizen, devised a shallow, covered, rectangular hearth for steel-making which later became known as the open-hearth process. Pig iron and scrap make up the charge which is heated by radiation from burning fuel gas just above the charge. Iron ore, preferably Fe_2O_3, is fed into the melt in controlled quantity to oxidize selectively the silicon and manganese and reduce the carbon level as desired. Although the first open-hearth furnaces were built with siliceous refractory lining, it was soon found that magnesia brick could be used. This strongly basic magnesium oxide allowed lime, that is calcium oxide, to be added substantially to remove phosphorus and sulfur impurities. The lime also combines with the silica and alumina. Cheap iron ores high in phosphorus and sulfur could now be used for making pig iron which was suitable for steel making *via* the open-hearth process. The first basic open-hearth furnace went into operation in the United States in 1888, and by 1890 there were 16 such installations at the Carnegie-Phipps Company in Pittsburgh, Pennsylvania. This large steel company undertook a major program in 1893–98 to replace Bessemer converters with open-hearth furnaces. Thus by the mid- to late 1890's, the low phosphorus content of Edison's concentrate ceased to be an important factor in the marketplace.

By early 1897 Edison knew that his fellow engineers had written off his "Ogden Baby" as a "monument of perseverance in original research." Where he had obtained $10 per ton for magnetite concentrates back in 1882 and had expected at least $6.50 for it from his present operations, a price as low as $4 was now non-competitive even with local iron producers. Despite the economic situation, he refused to give up until he had had an opportunity to operate his plant under "full throttle" and thus determine more quantitatively what his operating costs were. He was not one to stop a technical project before it was finished in order to cut his losses. Late in 1897 he accepted an order for 100,000 tons of concentrate from the Bethlehem Steel Company. The price appears to have been $3.50 per ton. Thus the loss of another hundred thousand dollars or so was not going to deter him from getting the data he had set out to get. He no doubt felt that the information obtained would be useful somehow, some place, and only he with his Ogden Baby was in a position to get it.

Edison made certain changes in his operations and made plans for volume production of 1,000 to 1,500 tons of concentrate daily. By autumn 1898 the 100,000 ton production run was underway. Late in November a severe blizzard forced the plant to shut down for several days. Edison was intent on getting all data possible from this prolonged run, and he wrote his wife that he was getting little sleep. He did not even leave the works to attend the funeral of his wife's sister in Ohio in November, nor did he return home for the Christmas holidays. Early in 1899 the Bethlehem Steel order was filled, and the plant was shut down. Although all stages of the operation had run reasonably satisfactorily, costs were much too high for 1899 prices. Prices continued to drop as Mesabi ore flooded the markets. Regardless of price, Edison would probably have been forced to give up at that location since the ore was then running only 10–12% iron, whereas the early diggings had contained 20–25%.

As indicated in the photographs, the New Jersey and Pennsylvania Concentrating Company was truly a commercial size plant. In the fall of 1898 when production was at a maximum, about 400 men were employed.

The Ogdensburg project was not Edison's final note in concentrating magnetite ores. The Dunderland Iron Ore Company was set up near Ranen Fjord, Norway in 1898, and it is known an Edison stone crusher was shipped for use in those operations. Theodore Lehman, who started working with Edison in 1888 and had been on leave from 1894–1898 to get his Ph.D. in Germany, was put in charge. Since magnetite makes up much of the primary ores available in Norway and many of them are high grade, it is likely the installation was used for concentrating lean ores rather than all of the ore production at the site. Whether the plant proved profitable is not known, but it ceased operation in 1908.

Costs of all three raw materials going into the blast-furnace operation are, of course, factors in the economics of preparing

Courtesy Edison National Historic Site, West Orange, N. J.

Edison in front of the foreman's office at the ore-concentrating plant

171

pig iron. In the 1890–1900 era, limestone was available at comparable costs in all of the iron-producing area. However, coke was not. The best coke for blast-furnace operations came from coal deposits near Pittsburgh. Coking coals are bituminous coals that readily melt, or fuse, in a retort or oven. Thus Edison's potential customers in the East were at a definite disadvantage here. Anthracite had shown possibilities for use in the blast furnace, and U.S. anthracite deposits are primarily in eastern Pennsylvania. In 1893 when blast-furnace operations were at a near standstill, Edison and the Crane Iron Works at Catasauqua, Pennsylvania carried out a joint program at the latter's plant trying to substitute anthracite for coke. Edison spent much time in Catasauqua during 1893. This work came to naught as did similar studies carried out by other iron producers, mainly because the anthracite broke up in the furnace before it burned.

All equipment used at the Ogdensburg site except the magnetic separators, the mixers for adding the binding agent, and the briquetting machines was used to manufacture cement (Chapter 6). The "Edison Crushing Roll Company" was set up to manufacture the rock-crushing machines, but this was handled on contract with an outside company. These machines were also sold to companies other than those controlled by Edison.

Anyone who studies the great skill shown by Edison in developing and commercializing home and office lighting following his invention of the first practical incandescent light is deeply puzzled as to how he got himself into such a sad situation in his ore concentration process. Tate attempts to explain Edison's poor judgment in taking up the concentration of magnetite on such a huge scale as his feverish desire to carve himself a large share of the iron and steel industry (18), the country's largest, in order to show those who had taken over his electrical interests in the formation of the General Electric

Company that he could outdo them. This hardly seems a logical explanation since Edison had obviously lost any desire to become a magnate in the electrical manufacturing field when he constructed his West Orange laboratories and plunged into several research projects long before General Electric was formed. Furthermore, he was active in research and development on concentrating magnetite in 1889 (*1*), three years before the G.E. merger.

The likely explanation for Edison's white elephant, or "Edison's Folly" as some of the newspapers were calling his ore concentration efforts in 1898, was his failure to do adequate market research before undertaking the grandiose project. The Mesabi discovery was made in 1890—*i.e.,* before construction had begun at the concentration works. Edison no doubt knew about the discovery shortly thereafter since he was an ardent reader of technical trade journals, and the Merritt brothers publicized their find since they were looking for financial support. Instead of investigating the find and potential future costs for bringing the ore to the Ohio-Pennsylvania markets, Edison chose to ignore the overall situation. As we saw in his refusal to consider alternating electric current as compared with direct current for lighting and power, he tended to push aside the work of others unless it was closely related to his own ideas. Edison certainly was a proponent of the strategy that the best defense is a good offense.

Edison's losses at the Ogden Mine would have been much less if he had designed it purely as a pilot or demonstration plant rather than as a combination pilot-commercial unit. For example, instead of 30 briquetting machines and 15 baking ovens to handle a total production of 1,500 tons a day, one unit of two machines and one oven would have given him adequate data as to costs of the agglomeration step. Since each of the first two stages of operations had large storage capacities for intermediate products, the entire plant did not have to run under comparable load from the quarry to the point of shipment.

Also, shutting down the plant early in 1897 following preparation of the 500 tons for the comprehensive blast-furnace test at Catasauqua would have saved a great deal of money and about two years in time. The price distress signals were as clear then as they were in 1899. Something more than logic must have kept him going. Several hundred families depended upon the operation of his concentration works for their livelihood. Many of his skilled mechanics had been brought in from a distance to work in this wilderness. Furthermore, he had invested his life's savings in the plant and its operations. He had spent more money, had a larger work force, and had worked a longer time on this project than he had on developing the incandescent light and demonstrating its utility. He had taken his young bride to their winter home in Fort Myers, Florida in 1886, and again in 1887, with the thought of returning there each year for a vacation. The Ogdensburg project caused postponement of a return visit until February of 1901. He had been forced to be away a great deal at a time when he had children at home. Edison had filed at least four patents a year since 1869, but he filed none in 1894–95 and only one in 1896. All this is ample evidence that he was very highly involved in something. Thus he and many of his employees had a lot at stake at the Ogden Mine, and perhaps if he could hold out for another year or two, the Mesabi ore development would prove disappointing.

Edison's contributions to the technology of mining, crushing, grinding, magnetic separation, and briquetting were all no doubt significant although no attempt has been made to evaluate them in detail. Magnetic separation of magnetite ores in New Jersey was reopened by the Alan Wood Steel Company in 1930 and continued at two locations until 1966 (19). At their mine near Washington, New Jersey, the original ore ran about

35% iron and was concentrated to 58–59%. At Scrub Oak Mine at Mine Hill, 25% ore was concentrated to 66%. At both of these mines, Edison's approach of concentrating all the ore was used. Production of the two mines was 250,000–300,000 tons of concentrate a year. The fines were shipped to the company plant at Conshohocken, Pennsylvania and agglomerated by sintering prior to feeding to blast furnaces. Thus, 30 years after Edison shut down, the price situation justified the concentration of magnetite ores but not starting with an ore as lean as that which Edison had attempted to use (*20*).

Edison met with some of his top operating people at the Ore Works early in the spring of 1899 to study the latest operating data, as well as the market situation. The decision was made to close the plant permanently, dismantle the buildings, and move or junk the machinery. W. S. Mallory, plant superintendent, related the following (*21*):

When this decision was reached Mr. Edison and I took the Jersey Central train from Edison, bound for Orange, and I did not look forward to the immediate future with any degree of confidence, as the concentrating plant was heavily in debt, without any early prospect of being able to pay off its indebtedness. On the train the matter of the future was discussed, and Mr. Edison said that, inasmuch as we had the knowledge gained from our experience in the concentrating problem, we must, if possible, apply it to some practical use, and at the same time we must work out some other plans by which we could make enough money to pay off the Concentrating Company's indebtedness, Mr. Edison stating most positively that no company with which he had personally been actively connected had ever failed to pay its debts, and he did not propose to have the Concentrating Company any exception.

In the discussion that followed he suggested several kinds of work which he had in his mind, and which might prove profitable. We figured carefully over the probabilities of financial returns from the Phonograph Works and other enterprises, and after discussing many plans, it was finally decided that we would

175

apply the knowledge we had gained in the concentrating plant by building a plant for manufacturing Portland cement, and that Mr. Edison would devote his attention to the developing of a storage battery which did not use lead and sulphuric acid. So these two lines of work were taken up by Mr. Edison with just as much enthusiasm and energy as is usual with him, the commercial failure of the concentrating plant seeming not to affect his spirits in any way. In fact, I have often been impressed strongly with the fact that, during the dark days of the concentrating problem, Mr. Edison's desire was very strong that the creditors of the Concentrating Works should be paid in full; and only once did I hear him make any reference to the financial loss which he himself made, and he then said: "As far as I am concerned, I can any time get a job at $75 per month as a telegrapher, and that will amply take care of all my personal requirements." As already stated, however, he started in with the maximum amount of enthusiasm and ambition, and in the course of about three years we succeeded in paying off all the indebtedness of the Concentrating Works, which amounted to several hundred thousand dollars.

Most mining towns become ghost towns after mining operations cease. The town of Edison and the surrounding area known as Summerville did more than that—they disappeared (22). At first the families at Edison refused to believe that the concentrating plant was permanently closed. However, when they began loading the machinery on freight cars, the workers knew it was all over. Fortunately, the New Jersey Zinc Company, with plants at both Franklin, New Jersey and Ogdensburg, which had been closed from 1880–1897 because of legal difficulties, resumed operations just about the time Edison was shutting down (23). This provided work for many of the miners and other personnel who were not retained by Edison for his operations at West Orange and elsewhere. Homes in the Edison-Summerville area were sold, dismantled in sections, loaded onto wagons and brought down the hill. A store constructed of two Edison houses today stands on Main Street in Ogdensburg, and there are two more nearby. Others which were moved as far away as Franklin still stand. The schoolhouse at Edison was also moved to Franklin and is now the Hungarian Church. The industrial buildings were made mostly of sheet metal and these were also dismantled. A stone building remained in Edison for a number of years. Later Charles Edison, son of Thomas and a former governor of New Jersey, had the building torn down and the stones moved to West Orange for use in constructing his home there.

I visited the Ogdensburg area in October 1969. The Edison site is about two miles up the hill from Ogdensburg on Edison Avenue. The "two million dollar hole" is now the municipal swimming pool. The surrounding area has many small valleys and streams. It is a beautiful spot. The area is now owned by the U.S. Steel Corporation and leased by the Ogdensburg Rod and Gun Club. The Ogden Mine Railroad which ran along the side of the mountain serving the various mines suspended operations in 1959, but the tracks remain in most places. As for the site of Edison's buildings, only pieces of briquettes and some evidence of the foundation of the power house remain.

The U.S. Steel Corporation has recently decided that their 2,000-acre property in the Sussex-Morris County area has greater potential as a residential and recreational real estate project than as a possible supplementary source of iron ore (*24*). Present plans call for the construction of approximately 5,000 dwelling units utilizing about half of the total acreage. The balance will be reserved for recreational use and open space.

Chapter 6 Edison's Venture in Cement and Concrete

THE MANUFACTURING AND MARKETING of cement was the only business operation which Thomas Edison ever entered which was already an established industry. Considering the fact that he had lost $2,000,000 ($9,000,000 based on 1970 dollars) in the iron-ore concentration project and had accumulated debts of several hundred thousands of dollars, one can certainly appreciate why he became a bit conservative for once in his life. However, taking into consideration the fact that the new products he introduced numbered several score, his high ratio of new products to total products is probably unequalled by any other diversified manufacturer, past or present. For example, many of the new products which present-day companies introduce are not actually new to the industry, but rather a matter of an individual company's carving out a share of an established market. Because of the arduous task of introducing a new product, many companies prefer to be second—*i.e.*, let the other fellow take the chances, bear the high introductory costs, and suffer the overall headaches which invariably accompany new product development. Although Edison entered a commodity market in going into portland cement manufacture, it was a relatively new industry in the United States, and little was known of the chemistry of the hydration reactions involved in the setting of cement. Edison played a major role in improving manufacturing operations and product quality. Furthermore, he spent about $100,000 in attempting to develop new methods for fabricating concrete structures.

178

Ever since man first started to build, he has felt the need for some cementing material to bind stones and other aggregates to form strong walls and floors and to give them a smooth appearance. The Assyrians and Babylonians are known to have used moistened clay for this purpose. The Egyptians used the mineral gypsum, which had been calcined or "burned," mixed with sand to form the mortar used in constructing the pyramids. Gypsum is calcium sulfate dihydrate. When heated in the 250°–270°F range, 1.5 moles of water per mole of gypsum are lost, resulting in the hemihydrate, $CaSO_4 \cdot \frac{1}{2} H_2O$. On mixing with a small quantity of water, the hemihydrate is converted slowly to the dihydrate, serving as a binder in the process. The hemihydrate is known as plaster of paris, having received its name from work carried out by French chemists in the late 18th century on the mechanism of the setting of calcined gypsum by reaction with water.

The Greeks produced lime by "burning" limestone and found that the action of the lime admixed with sand was enhanced if some volcanic ash was also used in the mixture. The Romans improved such mixtures by utilizing pozzolana, a volcanic deposit in Italy, which by chance had proportions of alumina and silica nearly optimum for use with lime to make an effective cement.

The quality of building cements declined with the decline of Rome, and the art of cement making was practically lost until the middle of the 18th century. It was not until 1824 that Joseph Aspdin, an English bricklayer, found that the addition of volcanic ash to the lime could be avoided if a limestone containing a relatively high proportion of clay was used and the calcination temperature carried to incipient fusion. The product was named portland cement because after hardening, it resembled a natural limestone quarried on the Isle of Portland, England.

Although the raw materials and the process for making portland cement are relatively simple, the chemistry of cement manufacture is actually highly complex. An example of the chemical composition of a high quality portland cement is as follows:

Lime	CaO	62.0%
Silica	SiO_2	22.0
Alumina	Al_2O_3	7.5
Magnesia	MgO	2.5
Iron oxide	Fe_2O_3	2.5
Sulfur trioxide	SO_3	1.5
Other	—	2.0

A cement can have such a chemical composition but still be worthless as a cementing material unless the basic chemicals are combined into the proper compounds. The major ingredients in a good cement are tricalcium silicate, dicalcium silicate, tricalcium aluminate, and tetracalcium aluminoferrate. After the calcination step, the magnesium oxide remains largely as such, there is some free lime, and the sulfur is present as calcium sulfate. The basic calcium oxide reacts mostly with the acidic silica and sulfur trioxide, and the amphoteric alumina and ferric (iron) oxide.

Tricalcium silicate is the essential ingredient in cement. The dicalcium salt hydrates much more slowly in the mortar than does the tri- but contributes greatly to ultimate maximum strength. The bonding properties of a cement in a sand–gravel–water aggregate are the result of gels, formed by the hydration of the silicates, plus reactions of the other ingredients with water. The anhydrous calcium sulfate is converted to gypsum as in plaster of paris, and free calcium and magnesium oxides are converted to their hydroxides as in regular plaster. The gels are largely colloidal, and their fibrous nature accounts in part for the strong bonding strength. It requires months for a concrete prepared from portland cement, sand, and crushed stone or gravel to reach maximum strength.

The two basic raw materials for manufacturing portland cement are limestone and clay; the former supplies the lime, and the latter, being mostly a mixture of aluminum silicates, supplies the alumina and silica. Many limestone deposits contain appreciable proportions of clay, such "clayey limestones" being known as cement rock. Since in such deposits the limestone and clay are intimately mixed, the early practice was to use cement rock as is if it had the approximate proper proportions of ingredients to make a good cement. The quarried rock was graded in size so that none of the stone was larger than about 12 inches in its longest dimension and none smaller than one inch. Vertical kilns, similar to those used to convert limestone to lime, were used for calcination. In making so-called natural cement, the temperature of calcination does not exceed about 1,800°F. The product obtained is unsintered, chalky, and porous, easily pulverized. At the turn of the century, most of the cement produced in the United States was of the natural type. Rotary kilns were also used to make natural cements, and the temperature was controlled to avoid sintering. Little natural cement is made in the United States today.

Portland cement is distinguished from natural cement in that it is calcined at 2,500°–3,000°F, which results in melting 20–30% of the reaction mixture to give a product consisting of hard clinkers. Since the partial melting results in better contact between the raw materials, it is easier to employ mixtures of raw materials rather than to rely on nature to use the right proportion of chemicals in the cement rock. Portland cement was first made in the United States in 1871 by David Saylor in Copley, Pennsylvania, using vertical kilns. The first rotary kiln was introduced in 1899, and within a decade the vertical kiln for portland cement became obsolete in the United States.

Thus, at the time the iron-ore concentration plant at Ogdensburg, New Jersey, was closed down, portland cement manufacture was still in its infancy. Edison had been interested in cement manufacture for some time and had several ideas as to how the operations could be improved. As to cement as a building material, he was quoted as saying, "Wood will rot, stone will chip and crumble, bricks disintegrate, but a cement and iron structure is apparently indestructible. Look at some of the old Roman baths. They are still as solid as when they were built." With such convictions, it is not surprising that Edison turned to cement manufacture as a means of utilizing equipment and know-how on mining, crushing, grinding, drying, and mixing which he had acquired in his abortive iron ore project. However, since cement making requires a large plant in order to be practical, it did mean a large capital investment at a time when Edison was in debt from his previous venture. He apparently arranged short term financing by mortgaging his holdings. In March 1904 the Edison Portland Cement Company sold $1,500,000 of 25-year 6% industrial bonds, the Williamsburgh Trust Company as trustee.

Cement rock had been discovered in 1898 in New Village, a small New Jersey hamlet (post office, Stewartsville), located about 45 miles west of West Orange. At that time it was served by both the Delaware, Lackawanna, and Western Railroad (DL&W) (1) and the Morris Canal. Beginning in 1906 a trolley line passed through the village from Phillipsburg to Washington, New Jersey. The Morris Canal, which ran from Phillipsburg to Jersey City, passed through both Stewartsville and New Village. It had been built primarily to move anthracite coal from the coal-rich Lehigh Valley, opposite Phillipsburg, to New Jersey and the New York City area via Lake Hopatcong, located in north central New Jersey. From an engineering standpoint, the Morris Canal was an amazing project in that the vertical rise or fall in a trip from the Delaware River to the Hudson

182

averaged about 18 feet per mile (2). In contrast, that of the Erie Canal from Lake Erie to the Hudson was less than one foot per mile.

After a survey by W. S. Mallory, former superintendent of the ore concentration works, Edison bought an 800-acre tract just south of New Village for his quarrying and plant operations. Since there was no limestone deposit at New Village, it was brought from Oxford, New Jersey, about five miles to the northeast. Edison built rail lines from the cement plant and the limestone quarry to the DL&W. This was called the Pohatcong Railroad, with original headquarters at the Edison Laboratories. Building of the plant began in 1899 and manufacturing in 1902. By 1905 production reached 3,000 barrels per day, making the Edison Portland Cement Company the fifth largest manufacturer of portland cement in the United States.

Edison himself designed the plant which included several unique features. The large mills from the ore plant were used to crush the quarried rock. The cement rock and limestone were ground to fine powders—97% through 100 mesh for feed to the kilns—by ball mills using both steel balls and rods. Whereas previously in cement mills the cement rock and limestone were proportioned with a wheelbarrow or other load units, Edison devised automatic scales with a hopper-closing device above each, electrically controlled by needles which dipped into mercury. Laboratory control chemists were responsible for adjusting the ratio of cement rock and limestone based on frequent analyses of the two feedstocks.

It was, however, the kiln itself where Edison made the most significant improvements. Let Mr. W. S. Mallory tell this story (3).

When Mr. Edison first decided to go into the cement business, it was on the basis of his crushing-rolls and air separation, and he had every expectation of installing duplicates of the kilns which were then in common use for burning cement. These kilns were usually made of boiler iron, riveted, and were about sixty feet long and six feet in diameter, and had a capacity of about two hundred barrels of cement clinker in twenty-four hours.

When the detail plans for our plant were being drawn, Mr. Edison and I figured over the coal capacity and coal economy of the sixty-foot kiln, and each time thought that both could be materially bettered. After having gone over this matter several times, he said: 'I believe I can make a kiln which will give an output of one thousand barrels in twenty-four hours.' Although I had then been closely associated with him for ten years and was accustomed to see him accomplish great things, I could not help feeling the improbability of his being able to jump into an old-established industry—as a novice—and start by improving the 'heart' of the production so as to increase its capacity 400 per cent. When I pressed him for an explanation, he was unable to give any definite reasons, except that he felt positive it could be done. In this connection let me say that very many times I have heard Mr. Edison make predictions as to what a certain mechanical device ought to do in the way of output and costs, when his statements did not seem to be even among the possibilities. Subsequently, after more or less experience, these predictions have been verified, and I cannot help coming to the conclusion that he has a faculty, not possessed by the average mortal, of intuitively and correctly sizing up mechanical and commercial possibilities.

But, returning to the kiln, Mr. Edison went to work immediately and very soon completed the design of a new type which was to be one hundred and fifty feet long and nine feet in diameter, made up in ten-foot sections of cast iron bolted together and arranged to be revolved on fifteen bearings. He had a wooden model made and studied it very carefully, through a series of experiments. These resulted so satisfactorily that this form was finally decided upon, and ultimately installed as part of the plant.

Well, for a year or so the kiln problem was a nightmare to me. When we started up the plant experimentally, and the long kiln was first put in operation, an output of about four hundred barrels in twenty-four hours was obtained. Mr. Edison was more than disappointed at this result. His terse comment on my report was: 'Rotten. Try it again.' When we became a little more familiar with the operation of the kiln we were able to get the output up to about five hundred and fifty barrels, and a little later to six hundred and fifty barrels per day. I would go down to Orange and report with a great deal of satisfaction the increase in output, but Mr. Edison would apparently be very much disappointed, and often said to me that the trouble was not with the kiln, but with our method of operating it; and he would reiterate his first statement that it would make one thousand barrels in twenty-four hours.

Each time I would return to the plant with the determination to increase the output if possible, and we did increase it to seven hundred and fifty, then to eight hundred and fifty barrels. Every time I reported these increases Mr. Edison would still be disappointed. I said to him several times that if he was so sure the kiln could turn out one thousand barrels in twenty-four hours we would be very glad to have him tell us how to do it, and that we would run it in any way he directed. He replied that he did not know what it was that kept the output down, but he was just as confident as ever that the kiln would make one thousand barrels per day, and that if he had time to work with and watch the kiln it would not take him long to find out the reasons why. He had made a number of suggestions throughout these various trials, however, and as we continued to operate, we learned additional points in handling and were able to get the output up to nine hundred barrels, then one thousand, and finally to over eleven hundred barrels per day, thus more than realizing the prediction made by Mr. Edison before even the plans were drawn. It is only fair to say, however, that prolonged experience has led us to the conclusion that the maximum economy in continuous operation of these kilns is obtained by working them at a little less than their maximum capacity.

The 150-foot kiln had a pitch of 18 inches, supported on rollers rotated by electric motors. Powdered coal, 80% through 200 mesh, was injected through tubes by air pressure from the lower end of the kiln in such a way that the combustion zone was extended over a length of 40 feet. The hot gases moving towards the feed end of the kiln were of such temperature that the limestone was converted to lime before reaching the combustion zone. Thus the by-product, carbon dioxide, did not interfere in the burning of the coal as was the situation in the smaller cement kilns. Air for the combustion was introduced in a spiral-type motion at the exit end to improve contact with the hot clinkers and thus reduce the heat lost in the hot product. Provision was made for longitudinal variations of the kiln resulting from thermal expansion. In 1926 the Edison plant had 10 of the 150-foot kilns. Edison's pioneering of the much longer rotary kilns proved to be an important technical contribution to cement making and similar calcination operations. Most of the cement trade subsequently licensed under his so-called "long cement kiln patents" (4), Edison receiving 1¢ per barrel royalty.

Edison used the same type of automatic oiling system for lubrication that he had used at his ore plant. The installation at the cement plant was at that time the largest such lubrication system in the country—10,000 bearings were lubricated in this way. Only two men per shift were required to handle the entire system for the one-half mile long plant.

The rate of setting and ultimate strength of a concrete aggregate depend on not only its chemical composition but also its particle size. Since nearly all of the chemical compounds in cement are water insoluble, the hydration reactions which the cement particles undergo are basically reactions at the surface of the particles. Since the surface of a sphere varies inversely as the cube of its diameter, the effect of particle size on the reactivity of a cement is readily apparent.

Edison experimented considerably on the effect of particle size on the setting properties of cements. He found that particles larger than about 200 mesh contributed practically nothing to original strength but would hydrate slowly after setting had

186

occurred, resulting in expansion and possible cracking. On the other hand, extremely fine material, such as that passing through a 300-mesh screen, reacted so rapidly with water that it would have little value in a concrete mortar or aggregate that was not immediately set in final form after mixing with water.

Edison considered the ideal cement as one which 100% passes through a 200-mesh screen but none through 300 mesh. However, he found that conventional grinding to a powder 90% through 200 mesh produced so much ultrafine material that setting was too rapid. He found that by carrying the grinding operation only partially to completion, separating and returning the coarse material, he could make a cement 85% of which passed through 200 mesh, and it still had satisfactory setting characteristics (5). He marketed such a cement which compared with material 75% through 200 mesh, characteristic of the trade at that time. It was the practice then as well as today to add 2–3% gypsum to cement to retard rapid setting. The action of the gypsum is primarily to control the rapid hydration of the tricalcium aluminate (6).

By joint action in 1917, the U.S. Bureau of Standards and the American Society for Testing Materials set up minimum quality standards for portland cement. It is likely that after such standards were accepted by the industry, Edison had little incentive to improve cement quality over that which met specifications. Since cement is used largely in contractual jobs such as paving and building which are obtained by competitive bidding, there is little or no incentive for the contractor to purchase a superior quality cement as long as a cheaper grade meets specifications.

Even though Edison entered the portland cement business at a time when there was severe competition from portland cement imported from England, and from lower cost domestic natural cement, his venture proved quite successful. One of the large projects in which Edison cement was used was the

Yankee stadium in New York. It was built in 1922–23 and required 180,000 bags of cement (94 lbs each). With a seating capacity of 67,000, it continues to be the second largest baseball stadium in the United States, with that of the Cleveland Indians No. 1. Edison was a leader in pioneering concrete highways. In 1905 an experimental one-mile stretch of concrete was built on Route 24 (now Route 57) near New Village. Edison furnished the cement and sand at no cost. Within a year the road had disintegrated. Rebuilt, the same thing happened twice more. This was finally attributed to the use of a clay base. A crushed stone base was used in building several miles near New Village about 1910. This road remained in use until 1953 when the old 18-foot wide pavement was used as a base for the present 24-foot road. Labor costs of the original paving were shared by state and county, and the materials were contributed by the Edison Portland Cement Company, Alpha Portland Cement Company, and Vulcanite Portland Cement Company. In 1950 Edward W. Kilpatrick, New Jersey State Highway Engineer, reported that cores taken from the 40-year-old pavement had tested 10,000 lbs per square inch in compression strength as compared with 4,500 when originally laid.

In November 1919 a fire in the cooperage shop of the Edison plant destroyed over 500,000 staves used in making barrels for cement. The fire raged for two days, and the Washington, New Jersey and Easton, Pennsylvania fire companies were called to supplement the Company's fire-fighting force. Total loss was estimated at $50,000, but the blaze did not seriously affect the plant operation since barrels were only used when packaging cement for export.

The fire engine which did yeoman duty at this fire was a 1906 horsedrawn steamer from Washington. A duplicate of this steam pumper now stands in the Edison Museum at Fort Myers, Florida. The original was contributed to the metal drive during World War II, but a duplicate was located by the former

Chief Engineer of the Washington Fire Department, Earle S. Eckel, Sr., after touring several states. Mr. Eckel was in charge of the American LaFrance fire engine during the 1909 fire and still resides in Washington.

About the time Edison's cement plant came into production, he started a limited research program on new methods for fabricating concrete structures. His primary objective in these studies might have been to increase the consumption of cement and thus benefit his manufacturing operations. Although this no doubt was a factor, it was very unlikely the dominating one. Edison never controlled more than a small fraction of total U.S. cement production. A pertient factor was undoubtedly because he considered concrete the building material of the future. He concentrated on one-family houses, and this has been attributed to his desire to see families from the slums relocated in such dwellings. When Edison travelled to New York City, he usually crossed the Hudson on the Hoboken-Christopher Street ferry. Christopher Street at that time was a wretched slum. Edison's frequent observation of the area motivated his search for methods to achieve low cost housing. As he himself said, "Invention should be able to remove millions of mothers and children from the human beehive tenements of the great cities."

Edison was actually a small-town boy at heart. He had moved his operations from Newark to Menlo Park, and later to West Orange, to be able to live and work in a rural community. There is no doubt that he felt that families living in one-family homes with lawns, trees, and flowers, would be a greater asset to America than families bunched together in city streets. This he believed could best be achieved with cast concrete houses.

Edison's work on the effect of particle size and chemical composition of a cement had shown that there were ways to control the setting rate. With his successful demonstration of various automated devices in his iron ore and cement operations, it is likely that he felt that house building was ripe for automation. Unfortunately, this goal remains today largely unattained.

Apparently Edison's concept from the start was to pour concrete in a single operation into molds making up the form of an entire house. This meant that the mixture had to have good flow characteristics—*i.e.*, it must remain uniform in composition while flowing both vertically and horizontally, and it must remain so until set to a solid composite. Such a mix had to be thinner than a typical concrete mixture in order to flow and to seek a common level as a liquid. Further, it had to contain a dispersing agent so that the coarse aggregate would not settle.

Certain colloidal dispersants, such as bentonite clay, impart to a liquid the property thixotropy—*i.e.*, the system is a firm jelly when standing, but movement causes it to liquify. As little as 0.85% of bentonite clay in water produces a soft gel structure. In about 1890 bentonite was used in drill muds in oil wells; the mud served to suspend the rock chips and bring them to the surface during boring. Edison was probably familiar with this development and used a similar approach in developing a concrete which could be poured.

Concrete mixtures were tested in the laboratory in a wooden mold of shape and dimensions indicated in the diagram (opposite). The test mixture was poured into the hopper until there was overflow at the other upright tubing. After the concrete set, the mold was dismantled, and the composition of the concrete was determined at various points. The concrete composition was standardized in parts by volume as: cement—1, sand—3, gravel (1/4–1/2 inch)—5. Use of 20% of a finely divided clay, based on the weight of cement, gave best results. Water was added to give sufficient fluidity so that the concrete

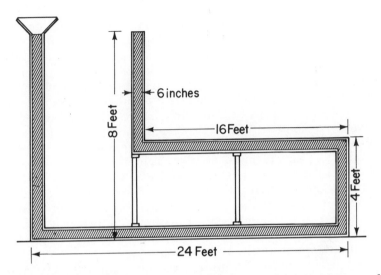

mix would flow under pressure. The clay not only imparted adequate stability to the concrete to pass the test as per the diagram but also improved surface characteristics. It was possible to adjust the cement composition so that the clay dispersant did not interfere in the rate of setting and had little effect on ultimate strength. With such a high proportion of coarse gravel in the concrete mix, it is indeed surprising that a uniform composition resulted after flowing a maximum of 56 feet as carried out in the test and standing for an hour or so before substantial setting resulted.

In his patent filed in August 1908, entitled "Process of Constructing Concrete Buildings" (7), Edison disclosed his approach:

The object of my invention is to construct a building of a cement mixture by a single molding operation—all its parts, including the sides, roofs, partitions, bath tubs, floors, etc., being formed of an integral mass of a cement mixture. This invention is applicable to buildings of any sort, but I contemplate its use particularly for the construction of dwellings, in which the stairs, mantels, ornamental ceilings and other interior decorations and fixtures may all be formed in the same molding operation and integral with the house itself.

In carrying out my invention I first construct a complete double wall house, which forms a mold for the reception of the cement mixture. This mold is preferably constructed of cast iron sections removably connected together in any suitable manner, as by means of bolts, dowels, etc., and adapted when the house is constructed and the cement mixture is hardened, to be taken to pieces and removed and used repeatedly for the construction of an indefinite number of houses. When the mold has been constructed and erected, I connect a number of distributing pipes therewith, which are preferably arranged at regular intervals at the top of the mold, the said pipes being connected to a common source of supply, which may conveniently consist of a vertical riser having a funnel-shaped opening for the reception of the cement mixture.

A Portland cement mixture especially adapted for this work is prepared adjacent to the mold and is raised continuously to the top of the mold and poured into the riser thereon, whence it flows around the top of the mold and is distributed evenly throughout the structure. An important feature of my invention consists in so regulating the rate of pouring the cement into the mold, taken in connection with the hard-setting time of the cement mixture, that the lower sections of the cement in the mold will have hardened before the level of the liquid cement above has risen very greatly, so that only a short column of the cement will act to create hydraulic pressure upon the mold. Because of this method, a much lighter and cheaper form of mold may be employed than would otherwise be necessary. It is also very desirable that the flow of cement into the molds should not be intermittent, as an intermittent pouring tends to produce seams. It is also desirable to have the mixture poured in as short a time as possible to save expense, but if the mixture is poured too rapidly, as above set forth, the molds need to be heavy to withstand the hydraulic pressure of the liquid cement.

Edison first worked on the molds for a two-family house which had been designed by a New York architect, but this was soon abandoned in favor of a one-family house. Two men worked in Edison's laboratory on the concrete house project for about eight years. One was Henry Harms who had worked on housing in the Far East possessions of the Netherlands; the other was George Small, a young engineer. The big problem was design of the molds, which constituted several hundred pieces for one house.

Charles Warner, editor of *Cement World,* visited the West Orange Laboratories in the spring of 1909 and interviewed Edison relative to his cement house, a subject which had caused considerable discussion and speculation in the press the previous two years (*8*). Warner was shown a model which had been built in the laboratory, he was given blueprints for all floors, and the entire building operation was discussed freely. Edison stated that the mold for the first house was about 75% complete, but this would be completed, and the first house would be poured before the end of the year (1909). A crane would raise the pieces of the mold in place, but all of the pieces for the inner walls would, of course, have to be light enough in weight so that they could be removed manually. The pieces of mold were locked together by flanges and backs reinforced by ribs. The bolts holding the inner and outer sections of the massive mold together passed through concrete collars which were left in the walls after setting. The small holes through the collars were substantially the only places in the concrete structure which had to be filled and smoothed over as a secondary operation. Floors, the roof, and other horizontal areas were reinforced with steel rods. Vents were provided in the mold structure to let entrapped air escape. These were flanged openings closed by an outer screen with an inner lining of filter cloth. All pipes, wires, etc. for the utilities were placed inside the mold with suitable terminals for connections. There were two rooms on each floor, and molds were designed so that either a two- or three-floor structure could be built. Only the windows

Photo by Stewart Fulton, Springfield, N. J.

A group of the cast concrete houses at Union, New Jersey. Photo taken in early fall 1970.

and doors were of wood and these were to be added after the mold had been removed.

In actual construction the concrete footing and basement floor were built before setting up the mold. The building operation would only be practical for a group of houses. Edison estimated that for efficient operation a builder should have six sets of molds. A crew of about 35 men could assemble the mold for a house in four days, it would require six hours to mix and pour the concrete, and the molds could be removed after four days. By staggering the operations in building a group of houses, Edison estimated the working force of 35 could construct about 144 houses in a year.

The Edison concrete house was not a commercial success although enough were built to prove that the basic process for the cast concrete house was technically feasible. The Edison experimental group built several such houses in West

194

Orange, including the garage and gardener's cottage at Edison's home, Glenmont. Edison never intended to enter the construction business; rather he wished to make his patents and know-how on the concrete house available to responsible builders on a royalty-free basis.

The most extensive housing project using Edison's technique was at Union, New Jersey, built by Charles A. Ingersoll, the famed dollar-watch manufacturer. Eleven houses were built and 10 on Ingersoll Terrace remain today, one having succumbed to the wreckers when the Garden State Parkway was built. These are all occupied by owners, Mrs. Joseph Fila having lived in hers for 49 years. Each house with a small lot origi-

Photo by Stewart Fulton

Mrs. Anne Connell standing in front of her
"Edison House" at Union, New Jersey

nally sold for $1,200. One recently sold for $18,000. Mrs. Fila maintains that the 24-inch walls keep out the summer heat and provide good winter insulation. Repairs are infrequent. Mrs. Anne Connell, whose family has occupied an "Edison house" for 18 years, asserts that the entire concrete structure is still in good condition. Her husband recently found that boring a hole through the wall to install a light fixture near the front porch was a "terrific job."

Ingersoll used only one set of molds, and the houses were all identical two-story construction. His workmen took much longer than expected to assemble the mold. Since small concrete mixers were used, it was difficult to coordinate the mixing with pouring, and considerable concrete was wasted. An unexpected stumbling block was buyer resistance. Whereas Ingersoll had expected his new houses to be snapped up by an eager public, apparently press publicity that such housing was for former slum dwellers gave them a stigma that a purchaser did not care to acquire (9).

Ingersoll's choice of 1917 for his building project would appear to have been an unfortunate one. At that time the United States was on a war economy, and labor was not only expensive, but scarce. The 11 houses were Ingersoll's last although he had not only planned 40 at the Union site but also a group using wooden molds at Monmouth Beach, New Jersey. Rumor was that Henry Ford was going to build 400 of the Edison houses near his River Rouge plant in Dearborn, Michigan, but this never materialized.

Edison had maintained from the start that his technique for building concrete houses would be practical only when carried out on a large scale at a given location. A composite mold was estimated to cost $25,000, and a minimum of six were required to keep a gang of men busy. In 1909 the crane, conveying and hoisting machinery, and mixer at the building site were estimated as an additional $25,000, making a total initial investment of $175,000—equivalent to $688,000 of 1970 dollars. Since non-industrial builders have tended to be individual business-men of limited means, there were probably extremely few with building experience who could consider going into such a venture. Edison estimated that his cast house on a small lot and including plumbing, heating, and light fixtures could be sold at a profit at a price of $1,200, assuming that sand and gravel were available near the building site. Certainly a good profit on a percentage basis would have to be realized on each house to pay the interest and depreciation on the high investment in molds and machinery.

In their report on concrete housing in July 1923, Hool & Johnson, Engineers, it was estimated that the completed house by the Edison process would cost $4,145 for labor and material—about $121 less than the cheapest frame-and-shingle construction of comparable size.* Quoting from this report:

Reduction or elimination of hand labor is the critical element in any housing operation. Mr. Edison showed rare courage and acumen in proposing to virtually wipe out finishing trades for houses; so revolutionary and so thoroughly economical a plan is sure to meet determined opposition from a number of sources.

Opposition by the building trades was probably a major deterrent in the commercialization of the Edison concrete house.

A staff article entitled "Review of Edison's Concrete House" appeared in 1965 in a publication dealing with concrete (*10*). This article contended that the Edison mold for a single housing unit constituted 500 pieces. Assembling such a mold was too complicated, especially for the workman who was not interested in making the system work anyway. The article states ". . . even for Edison, the organization of the molds was a

---

* A copy of the Hool-Johnson report is on file at the Edison National Historic Site. It was apparently not published. The Consumer Price Index was 59.4 in 1923 compared with 32 in 1909.

feat not easily solved." Conclusions were that "Edison was ahead of his time. His monolithic concrete houses never became popular."

Another innovation relating to concrete construction which had its beginning in the Edison Laboratories was foam, or cellular, concrete. A patent issued to Aylsworth and Dyer of Edison's staff in 1914 is the basic patent on foam concrete (11). The product, called porous artificial stone by the patentees, was prepared by adding 0.25–1.0% of fine aluminum powder to an alkaline hydraulic cement. Aluminum reacts with alkalies to form hydrogen, which in this case served to create porosity as the cement mortar set.

Foam concrete has become an important commercial product for use in building construction where insulation, and not maximum strength, is desired. Although the technique of generating hydrogen *in situ* with aluminum powder continued to be used until recently, the trend now is to use a foam concentrate with the cement–water mixture. Without filler with the cement, the foamed product weighs less than 50 lbs per cu ft (water—62.4 lbs). Such lightweight concrete finds its major use as roofing. For walls or flooring, sand is included in the mixture to be foamed, to give a composite of about 90–100 lbs per cu ft. According to the Hool-Johnson report mentioned above, Edison did some work with foam concrete in connection with his concrete house development.

Early in 1970 the author contacted the U.S. Department of Housing and Urban Development to determine if the Edison approach was being considered in their Operation BREAKTHROUGH, a program directed to overcoming this country's housing deficit. Apparently Edison's cast concrete house was not proposed by any member of private industry although several proposals dealing with precast concrete were received. HUD did not consider Edison's technique competitive with the precast approach, mainly because of higher labor costs at the site *vs.* prefabrication in factories (*12*). An example of a precast method is that used recently by Dr. George J. Kreier, Jr. of Philadelphia, to build 5,000 low cost homes in Jamaica (*13*).

A factor which has probably prevented the Edison technique of pouring an entire house from receiving attention in recent times is "the-not-invented-here" resistant attitude of most research and development personnel. Considering the fact that the Edison house has been amply demonstrated by the 50-year field test of the 10 houses at Union, New Jersey, and with the public's placing a present value on such houses well above projected cost,* it is difficult to see why Edison's approach should not have possibilities for current low cost housing. Edison's big problem was with his cast iron molds—their heavy weight, large number of pieces, and the need to plate the inner surface with brass or nickel to obtain good release properties. Use of fiberglass-reinforced thermoset plastic for mold construction would reduce the overall weight of the composite mold about 75%, allowing fewer pieces and at lower cost. Admittedly, construction by the Edison technique in an area with less rigid building codes than in the United States and with labor at the building site available at prices competitive with the national average would increase the likelihood of success. By using fiberglass–plastic molds tooling costs would be relatively low, and the manufacture of the molds would be less sophisticated than if they were made of metal. Because of the relatively low strength required in a two-story house, it is likely that natural cement could be used, a material available in some developing countries where portland cement is not.

Edison's interest in cement led him to research in still another direction—furniture. Rumors in 1911 that he was working on concrete furniture reached the trade, and he was

---

* Based on estimates by both Edison and the Hool-Johnson report, adjusted to 1970 dollars in accordance with Consumer Price Index values.

interviewed by a representative of *Furniture World* on the subject (*14*). The project was apparently well underway since the visitor was shown a phonograph cabinet which was "as smooth as satin." Construction of a cement piano was being considered. Edison commented that his cement furniture was only 55% heavier than comparable wooden furniture and that he expected to get this down to 25%. Obviously he was using foam concrete, possibly reinforced. Edison said that he had shipping tests of furniture in progress to Chicago and New Orleans. He predicted that it would be "possible for the laboring man to put furniture in his home more artistic and more durable than is now found in the most palatial residence in Paris or along the Rhine."

This project also was not commercially successful, and it seems that Edison did not push it—possibly because of breakage encountered in the shipping tests. Regardless of the reasons for failure, one must admit that this was an innovation that few would attempt. Imagine what Edison could have done if he had had the plastics available today.

Edison's venture into cement manufacture was quite successful for the first 30 years or so. In November 1919, Superintendent A. H. Moses reported that the plant was running at full capacity and had on hand orders that would keep it running for months to come. Plant capacity was about 10,000 barrels (3,760,000 lbs) per day, with about 500 men employed. However, the supply of high quality rock became a limiting factor in the early 1930's. Exploration showed that there were substantial deposits underground, but shaft mining could not compete with open-pit quarrying. Once the local cement rock could not be used to a maximum, the plant became non-competitive. In 1937 it was closed down and in 1942 sold to the Chilean Import Company who bought the property

Courtesy Edison National Historic Site,
West Orange, N. J.

Phonograph cabinets constructed from cement

primarily for its electric plant which was dismantled and shipped to Chile. Over the years the buildings have been occupied by a series of small companies. At present the largest installation at the former Edison site is the Elizabethtown Gas Company, formerly the Northwest Jersey Natural Gas, Incorporated. The Elizabethtown Gas installation does not occupy any of the former Edison buildings.

Edison probably would never have entered into cement manufacturing if it had not been a means of using equipment and know-how which he had accumulated during his iron ore concentration project. It no doubt was a wise decision and served to help recoup his earlier financial losses. However, once the plant was in successful operation, he largely lost interest in

201

Courtesy Edison National Historic Site,
West Orange, N. J.

Edison as he appeared when making a periodic inspection of his cement plant

it. Unlike his close associates Henry Ford and Harvey Firestone, Edison was not interested in passing on an industrial empire to his children. Neither of his sons had strong interests in manufacturing. Thus, six years after Edison's death, Edison cement was no more. There are still several former employees of the Edison Portland Cement Company living in the New Village area. Since 1964 they have had a reunion each fall.

Chapter 7 The Amazing Alkaline Storage Battery

THE GAY NINETIES got its name largely from the fact that it was the decade of the bicycle. Interspersed with carriages and electric trolleys on city streets were the individually controlled tall-wheeled "ordinaries," with their trailing diminutive second wheels, and the newer "safeties," with two small wheels. Bold men in knickerbockers, as well as daring ladies in bloomers, took to the bicycle in ever increasing numbers. The new-fangled pneumatic tire gave cycling a tremendous boost, as did the surfacing of streets with stone or asphalt.

The bicycle as well as the carriage served as a prototype for the self-propelled vehicles which appeared on the scene in numbers at the turn of the century. Ford's first "auto" built in 1896 was called the quadricycle. Steamers, gas buggies, and electric motor-storage battery driven cars vied for the dollars of the venturesome customer. Initially the quiet electric cars were more popular than the noisy, hot, and troublesome gasoline driven vehicles. A survey made in New York in late 1899 showed that of 100 motorcabs in the downtown area, 90 were powered by storage batteries.

Edison liked this new form of transportation and allowed that the horse was doomed. He saw advantages for the electrically driven car, particularly if a battery superior to the lead-acid type could be developed. He commented to the effect that if Nature had intended man to use lead in batteries for powering vehicles, she would not have made it so heavy.

The term "battery," as commonly used in electricity and electrochemistry, is a device for converting chemical energy directly to electrical energy. The chemical reactions within a battery cell cause electrons to be released at one electrode, which, in a closed circuit, flow through an external conductor to the other electrode. As electrons move from one electrode to another by an external circuit, ions within the cell migrate so as to neutralize the effect of the external movement. Electric cells are classified as primary or secondary. The early type were all primary cells. Those used in early telegraphy are described briefly in Chapter 1.

Any electrochemical cell is made up of two half-cells. Each half-cell consists of an electrode, which is a metal capable of conducting electrons, and an electrolyte, which in most cases is a water solution of an acid, base, or salt. The positive electrode is the one which acts as an oxidizing agent; that being oxidized is the negative electrode. A secondary cell, also called a storage cell or accumulator, is capable of being recharged and reused as a source of electric power. During the recharging, the electrodes, by definition, are reversed. The two types of electrodes are also defined as anode and cathode. However, since this terminology is not uniformly defined by writers on the subject, it is not used here.

During the 19th century experimenters discovered that when a direct current passed between certain pairs of electrodes of the same metal immersed in an electrolyte, the electrodes became polarized—i.e., if they were connected by an external circuit, a current flowed. In 1859 Gaston Planté, a Frenchman, studied the system of lead electrodes in dilute sulfuric acid. He found that a coating of lead peroxide was produced on the positive plate. Reversing the flow of current several times in the charging operations converted the plates to a spongy condition with increased activity. Primary cells were the only means of charging, requiring months to convert a pair of lead plates to an active storage cell. In 1881, Faure in France and Brush in America applied lead oxide to lead plates and charged them to produce a storage battery; the positive electrode became coated with lead dioxide, the negative with spongy lead. Subequent work led to the use of netlike lead grid structures for holding the active ingredients. Reduced formation time, use of highly efficient electric dynamos for charging, and the demand for high capacity batteries for the mechanical transportation field all led to rapid commercial development of the lead–acid battery by the end of the century.

When any metal is immersed in an acid solution, it tends to dissolve, giving up electrons to the mass of the metal. In the case of the lead cell:

$$Pb \rightleftharpoons Pb^{2+} + 2\,e$$

If the lead electrode is connected to the lead peroxide electrode by an external conductor, the electrons leave the lead, shifting the above equilibrium reaction to the right. The electrons reduce the four-valent lead (as lead peroxide) to the bivalent form. Thus the net reaction in the sulfuric acid electrolyte is:

$$Pb + PbO_2 + 2\,H_2SO_4 \rightarrow 2\,PbSO_4 + 2\,H_2O$$

The lead sulfate is highly insoluble in the dilute sulfuric acid, replacing the spongy lead and lead peroxide on the grids. One can see from the above chemical reaction why specific gravity can be used to determine the approximate state of a lead storage battery.

Recharging the battery with a flow of electrons in the opposite direction results in reduction of bivalent lead to metallic lead at the one electrode and oxidation of the other by the electrolyte.

Reduction: $$PbSO_4 + 2\,e \rightarrow Pb + SO_4^{2-}$$

Oxidation: $$PbSO_4 + SO_4^{2-} + 2\,H_2O \rightarrow PbO_2 + 2\,H_2SO_4$$

The negative ion, SO_4^{2-}, migrates from the lead electrode to the other electrode and provides an atom of oxygen from a molecule of water.

As other electrical scientists, Edison had used lead batteries and found them useful reservoirs of power. His experiments with them indicated that a better storage battery should be possible, and, starting in 1900, he undertook a major laboratory program to this end. He appeared to have two prerequisites: the new battery would not contain lead, and the electrolyte would be alkaline.

At the time, Edison was manufacturing the primary battery known as the Lalande cell. This battery has zinc and copper oxide as electrodes in strong sodium hydroxide solution. The cupric oxide is compressed into a rectangular copper-plated steel frame held together by a suitable binder. The zinc electrode is high purity zinc containing 0.5–2.5% mercury. The sodium hydroxide electrolyte is about 6 normal (240 grams of NaOH per liter of solution). The zinc, being amphoteric, tends to dissolve, forming sodium zincate (Na_2ZnO_2) and leaving excess electrons in the metal. When the electrodes are connected by an external circuit, electrons flow from the zinc to the cupric oxide electrode, reducing the latter to cuprous oxide and then to metallic copper. Such a primary cell has excellent shelf life, a relatively high ampere-hour capacity, and constant voltage discharge.

At the time Edison started experimental work on the storage battery, he was manufacturing several hundred thousand Lalande primary batteries annually for the railroads of the country for signal service (1). Thus, he used his know-how on this primary battery in his first research on storage cells. Use of zinc as an electrode was out since it forms a soluble compound in an alkaline solution. In a storage battery the No. 1 prerequisite is that the oxidized and reduced forms of the materials making up the electrodes be insoluble in the electrolyte and stay put. Thus, in a storage battery the active ingredients of the electrodes must be finely divided, or spongy as in the case of lead, so that the electrolyte can penetrate throughout the electrode and make possible a much higher capacity than if only the surface of the electrode reacted.

Edison first studied the copper oxide electrode of the Lalande cell to see if it could be made reversible for storage battery use. This approach was not successful. Copper is next to the base metals—mercury, silver, platinum, and gold—in electrical potential and cannot be oxidized under conditions used in storage battery charging. Furthermore, the lower valent oxide, Cu_2O, is an oxidizing agent as well as the cupric form, CuO, and thus it is not possible to stop at the cuprous stage during discharge.

Porous carbon strips, their pores filled with test material, were screened for the positive electrode, coupled with a standard negative half-cell. Test tube size electrodes were used, and capacity in milliampere-hours was the primary criterion of quality. Even though the reaction at an electrode could be reversed and the chemical substance restored to its original state by charging, the electrode was of no interest unless it was a good source of energy per unit weight. Once a positive element looked promising it was checked out against possible negative electrodes.

After a great deal of screening, Edison concluded that nickel oxide was best for the positive electrode and iron for the negative. He also found cadmium to be a promising material for the negative half-cell and in 1900 obtained a British patent on the nickel–cadmium couple. Kirk-Othmer reports as follows (2):

Edison was aware of the lower self-discharge and greater charge efficiency of cadmium, but decided against its use, in favor of iron, for two reasons: (a) Edison was seeking a fool-proof battery which would deliver the highest possible number of deep discharge cycles; this is the fundamental requirement of the vehicle propulsion application which he expected would become very important as electric power became generally available in the United States. The iron anode appeared to be eminently satisfactory in this respect. (b) Cadmium was and still is a scarce metal; there is only twice as much cadmium as silver in the earth's crust. Edison did not foresee the uncovering of sufficient ore deposits to fill the battery production volume he anticipated.

Cadmium was selling for $4 per lb at the time.

Waldemar Jungner, a Swedish chemist, was also working on storage cells about the same time as was Edison. Jungner's work led to the nickel–cadmium storage battery which is today a familiar alkaline type. He had little interest in its potential use for vehicle propulsion; rather his objective was to develop a battery capable of a variety of application conditions. Jungner obtained Swedish patents on the nickel–cadmium system in 1899–1901. He was joined by a fellow Swede named Berg, and factory production of their battery began in 1909–10.

Edison found an approximately 25% potassium hydroxide solution to be the preferred electrolyte for his nickel–iron system. A major advantage of this base over sodium hydroxide is its greater solubility in water at low temperatures and thus lower tendency for the concentrated solution to gel in winter use. Actually sodium hydroxide was used in the Edison battery during World War I when the supply of potassium salts from Germany, then the primary source, was shut off.

The reaction in the nickel–iron battery can be simplified to the following:

$$8 \text{ NiO(OH)} + 3 \text{ Fe} + 4 \text{ H}_2\text{O} \rightleftarrows 8 \text{ Ni(OH)}_2 + \text{Fe}_3\text{O}_4$$

$$\rightarrow \text{Discharge}$$

$$\leftarrow \text{Charge}$$

Unlike the lead–acid battery, the electrolyte undergoes no net change either during discharging or charging. As indicated in the above equation, the nickel is primarily in the trivalent state after charging. However, a freshly charged battery appears to contain an appreciable amount of a higher valent nickel oxide. Nickel dioxide, and possibly higher nickel oxides, appear to exist only in a transitory state. During discharge, the iron is likely first oxidized to ferrous hydroxide, which is further oxidized or breaks down to Fe_3O_4. The latter is a true oxide and not a mixed ferro-ferric type.

Early in the development work on the two electrodes Edison made the wise decision to prepare, rather than purchase, the chemicals going into the battery. A chemical works was set up at Silver Lake—an area below Bloomfield—about five miles north of the laboratory area. It was first a pilot plant for preparing experimental batches of materials for testing, but later the manufacture of chemicals for the battery became an integral part of the manufacturing operations. A separate laboratory for battery research was built at nearby Glen Ridge. Work here supplemented the more exploratory studies at West Orange.

When Edison began his battery work, high quality nickel was not commercially available. He was forced to use a nickel-bearing alloy which he purchased in the form of shot. The first step in preparing pure nickel hydroxide was to dissolve the shot in hot concentrated sulfuric acid. The hydrogen evolved was saved. Any copper impurity was precipitated by the electrolytic action of the nickel. Arsenic and antimony were precipitated by hydrogen sulfide treatment, and the sulfides were filtered out. The iron impurity was oxidized to the ferric stage by sodium hypochlorite, the solution neutralized with sodium carbonate, which precipitated the iron as ferric hydroxide. The purified nickel sulfate filtrate was then sprayed into boiling 20% sodium hydroxide solution in a copper-lined tank. The $Ni(OH)_2$ precipitate was filtered, washed, dried, rewashed, and redried. It was then ground and screened to 30–190 mesh size for battery use (3). In 1918 the International Nickel Company made available a high purity grade of nickel suitable for use in batteries.

Iron for the negative cell was prepared from iron powder of very low manganese content, obtained initially from a Swedish source and later from Armco; this was dissolved in dilute sulfuric acid at ordinary temperature in a lead-lined tank, and the hydrogen was saved. The ferrous sulfate obtained was purified by crystallization from water (as $FeSO_4 \cdot 7 H_2O$), centrifuged, and finally dried at 200°C. During the drying operation, the heptahydrate would be converted to the monohydrate, $FeSO_4 \cdot H_2O$. Edison then oxidized the dried sulfate to the ferric oxide in a muffle furnace. The Fe_2O_3 was leached with sodium hydroxide solution, washed with water, dried, and reduced to iron using mainly the hydrogen recovered from the two dissolution processes. The hydrogen was purified by bubbling it through sodium hydroxide solution and chromic acid solution prior to use.

Edison was probably the first to manufacture iron powder by reducing an iron oxide with molecular hydrogen. He used a fixed-bed technique—*i.e.*, ferric oxide was placed in iron retorts in a muffle furnace at a temperature of at least 1200°F, and the hydrogen passed through the retorts at near atmospheric pressure. Iron powder so prepared is much more chemically active than that prepared mechanically from molten iron by atomization. A commercial process was recently developed to reduce iron ore by the fluid-bed technique; it uses gaseous reducing agents such as hydrogen and carbon monoxide. Iron is recovered as a solid powder—no liquid product is formed. Iron powder is also prepared by the thermal decomposition of iron pentacarbonyl vapor. None of these latter processes was available at the time Edison was pioneering several new chemical operations at his Silver Lake plant. He supplied iron powder similar to that used in his battery to reagent suppliers as pharmaceutical-grade iron. The present manufacturer of the nickel–iron battery continues to do this. The only grade that is purer is that prepared by gaseous decomposition reactions such as that of iron carbonyl.

In Edison's process the reduced iron was cooled under an atmosphere of hydrogen and then stored under a layer of aqueous potassium hydroxide. Prior to use it was dried, ground to break up any agglomerates, and mixed with 3% of the yellow oxide of mercury, HgO. The latter was reduced to metallic mercury when the battery was subsequently charged. The fact that an approximately 40% stoichiometric excess of iron was used in the negative cell provided adequate conductivity when the active iron was in the oxidized state. Crystallographic data indicate that the iron alkaline electrode is initially oxidized to ferrous hydroxide, $Fe(OH)_2$, but on more complete discharge the tetroxide, Fe_3O_4, is formed.

The ingredient powders of both electrodes were originally packed in pockets made from cold rolled carbon-steel ribbon which were perforated when they passed through special rolls which punched 560 holes per square inch. The burrs were removed by emery wheels, and the ribbon was cleaned by revolving steel brushes. Then as a continuous process, the ribbon passed through a caustic soda cleaning bath, a tank of hot water, and the nickel-plating bath consisting of nickel sulfate–nickel chloride. The nickel-plated ribbon was washed with water and with dilute ammonia and annealed in hydrogen.

The pockets used to hold the active ingredients were made from the nickel-plated ribbon cut to length and pressed into a form similar to two halves of a pasteboard box. These halves were put together, leaving one end for filling. The pockets were fastened in a nickel-plated steel grid under hydraulic pressure which sealed the boxes and crimped the sides. The pockets were $3 \times \frac{1}{2} \times \frac{1}{8}$ inch, and 24 fitted into a grid.

The necessary number of plates of each kind were mounted on steel rods which passed through the eyes of the grid at the top of the plate. The groups of positive and negative plates separated by hard rubber pins were intermeshed to form the cell. Hard rubber was also used in sheet form to insulate the cells from the sides of the nickel-plated steel container. Although the pockets containing the active electrode ingredients were small and metallic, the nickel hydrates were such poor conductors that the internal resistance of the battery was much too high to be practical. After much searching it was decided that a flake-type natural graphite admixed with the nickel hydroxide solved the problem.

The factory at Silver Lake was enlarged, and manufacture of the new battery began in 1904 with considerable fanfare in the newspapers. The battery found use in small industrial trucks which had come into service about 1900 and automobiles such as those of the Studebaker Brothers Manufacturing Company of South Bend, Indiana. Edison had his sights primarily on the electric automobile, and durability tests were initiated over rough roads, which were quite commonplace in those days.

Within about six months after the battery had been introduced, it was evident that many of those in service showed losses in power up to 30%. There were other defects, such as leaking containers and overall uneven performance. Despite pleas from many in industry to continue production of the battery, in 1905 Edison decided to shut down manufacturing operations. Purchasers were refunded full price for any defective batteries returned. It was a severe blow, especially since sales had been started abroad as well as in the United States. Edison had borrowed $500,000 on his holdings to start the Edison Storage Battery Company venture, and he had sunk over $1,000,000 into research and development on the battery.

The average manufacturer would probably have stopped work on the battery at this point. After all, irrespective of the defects, the battery proved to be less efficient than the lead–acid battery—*i.e.*, a lower percentage of the electrical energy charged was recovered on discharge. The voltage of the battery was disappointingly low. Furthermore, there was no criterion that any successful type of alkaline storage battery was possible.

However, Edison gave no thought to abandoning his battery project. Following his philosophy that there is a solution to every problem if one works hard enough, research on the battery was resumed on a pattern reminiscent of the Menlo Park days when work-work-work was the way of life.

Defective batteries which had been returned were dismantled, and all features were scrutinized for possible improvements. It was obvious that the No. 1 problem was with the positive plate. The conductivity of the nickel hydrate pockets had declined with use, despite the presence of the conductive graphite. Graphite has a specific gravity of only 2.25 as compared with 4.1–4.4 for the nickel oxide hydrates. Successive swelling and contraction of the nickel hydrates during use resulted in a non-uniform dispersion of the graphite and densification of the nickel in the bottom of the pockets. However, the most serious defect of the positive electrode was that swelling of the nickel hydroxide on oxidation appeared to be somewhat cumulative, resulting in stresses beyond the elastic limits of the perforated steel containers and disarrangement of the containers in the grid.

The successful solution to the problems associated with the positive electrode is ample justification for calling Edison's alkaline battery "amazing" not only because of the improved electrode, but also because of the intricate machinery and processes which were developed to manufacture it.

Edison favored nickel as a conductor for the nickel oxide since it was not attacked by the electrolyte, was resistant to oxidative attack, and was chemically similar to the active electrode. However, the difference in the specific gravity of nickel *vs.* the nickel oxide hydrates (8.9 *vs.* 4.1–4.4) was even greater than with the graphite, which ruled out the use of finely divided powder. Considerable work was done with nickel wire and foil. Success came with the development of nickel flake 0.00004 inch thick (4).

The manufacture of this nickel flake was indeed an intricate and expensive operation. Ten revolving copper cylinders were carried by a crane, dipped alternately into copper- and nickel-plating baths, and sprayed with water after each plating. This process of alternate plating was repeated 125 times in a five-hour period. Later the process involved the laying down of 150 layers of each metal. The 0.0075-inch thick copper–nickel composite consisting of 250 layers was then stripped from the copper cylinders and cut into 1/16-inch squares. The copper was then dissolved chemically, first with aqueous ammonia containing a mild oxidant but later with a solution of sulfuric acid saturated with nickel sulfate. The thin flakes were washed, centrifuged, and dried over steam coils. This flake nickel was so thin it would float in air like thistledown. A bushel weighed only $4\frac{1}{2}$ lbs despite the fact that nickel is 8.9 times heavier than water.

The improved containers for the positive electrode were cylinders, $\frac{1}{4}$ inch in diameter and $4\frac{1}{8}$ inches long, prepared by spiral winding of the perforated nickel-plated steel ribbon. The seams were lapped and sealed by pressure. The tubes were made by spiral winding, both clockwise and counterclockwise, and by alternating the two types in the assembly to equalize strains that might cause buckling. A cap was placed at the bottom of each tube, and a group of eight was placed in molds of a machine designed to fill and tamp the active materials in place. The machine had two hoppers, one for the nickel hydroxide and the other for the flake nickel. A charge of each would be added to the eight tubes, and then weighted ramrods would fall by gravity into the tubes striking a blow of 2,000 lbs per sq inch. This process was repeated 300 times; it resulted in 300 layers of the hydroxide of about 0.01-inch thickness sepa-

Courtesy Edison National Historic Site,
West Orange, N. J.

This photo was taken in 1906—a time of maximum research effort
on the storage battery.

rated by layers of the flake nickel 0.001 inch thick. These layers
made up 12% of the tube contents, the total weight in each tube
amounting to 9.20 grams. When filled, the ends of the tubes
were pinched off, and each was reinforced with eight seamless
rings spaced equally around the tube. A total of 30 tubes fitted
into a grid, 15 in both the lower and upper sections.

The negative half-cell was not changed in the improved battery. The electrolyte was modified in that the potassium hydrox-

ide concentration was reduced from 25 to 21%, and 50 grams per liter of lithium hydroxide were added. Since the lithium hydroxide tended to concentrate in the positive tubes, the renewal electrolyte consisted of 25% potassium hydroxide and only 20 grams of lithium hydroxide per liter. However, to ensure a saturated solution of the lithium, a few grams of the lithium hydroxide monohydrate were placed in each cell before charging the electrolyte. Although it is an alkali metal, lithium compounds in many respects resemble those of the alkaline earths. The solubility of lithium hydroxide is much lower than that of sodium, potassium, rubidium, or cesium. The lithium has a protective effect on the positive electrode, resulting in longer battery life. The low atomic weight of lithium, 6.94, as compared with 39.1 for potassium, with their resultant differences in atomic diameter, are major factors in the behavior of the lithium. A layer of lithium hydroxide molecules appears to form on the nickel oxide hydrate crystals. However, there is still no proved theoretical explanation for the beneficial effects obtained with the lithium.

Edison had to build his own plant to manufacture the hard rubber pins used as insulation between the grids and the sheets on the sides of the container. The complete assembly of plates rested on a bridgework of hard rubber at the bottom, shaped to hold the plates in proper position. Hard rubber contains 35–40% sulfur whereas regular vulcanized rubber contains less than 5%. Hard rubber is a plastic of substantially no elasticity, and its use resulted in severe foaming of the electrolyte. This was traced to soaps on the surface of the rubber from mold-release agents. The Edison Storage Battery group probably used glycerin as a mold release and washed this water-soluble material from the surface of the hard rubber prior to use.

The container for assembling positive and negative plates with electrolyte and insulation was a nickel-plated steel can. The side seams and bottom were welded, and the joints were

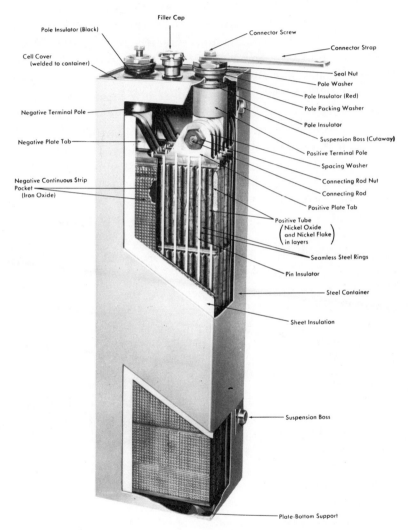

Pole Insulator (Black)

Filler Cap

Connector Screw

Connector Strap

Cell Cover
(welded to container)

Seal Nut

Pole Washer

Pole Insulator (Red)

Pole Packing Washer

Negative Terminal Pole

Pole Insulator

Suspension Boss (Cutaway)

Negative Plate Tab

Positive Terminal Pole

Spacing Washer

Negative Continuous Strip
Pocket
(Iron Oxide)

Connecting Rod Nut

Connecting Rod

Positive Plate Tab

Positive Tube
(Nickel Oxide
and Nickel Flake
in layers)

Seamless Steel Rings

Pin Insulator

Steel Container

Sheet Insulation

Suspension Boss

Plate-Bottom Support

Courtesy Alkaline Battery Division, ESB, Inc.

Diagrammatic sketch of the improved Edison battery

again nickel plated. The details of the assembly, recommended charging and discharging conditions, care of battery, and so on, are available from various treatises on storage batteries (5).

Edison resumed manufacture of the nickel–iron alkaline battery early in 1909, and manufacture has continued to date. Originally four, six, and eight positive plates—the number of negative plates was always one more than the positive—were used per cell. A battery could consist of as many cells as desired. A 30-cell, eight positive-plate battery is common for powering industrial trucks. Edison certainly achieved his objective of developing a more durable battery, both from the standpoint of handling and abuse. His alkaline battery also provided more electrical energy per unit weight of battery but not to the degree which he originally foresaw.

Holland, an employee of Edison's, reported in 1910 that the Edison battery gave the most power per unit weight of any battery on the market, but he gave no specific data (4). Salom, who investigated the lead–acid battery, in 1910 said that the Edison cell gave 15 watt-hours per lb *vs.* 10 watt-hours for the lead–acid (6). Two Cornell University investigators reported in 1913 that for equal amounts of power, the Edison battery weighed 25% less and cost 25% more than an ironclad lead battery (7). In 1924 the Edison Storage Battery Company asserted that the Edison battery was 20–25% lighter per unit of power than the best lead–acid battery and had twice the service under similar conditions (8). The lead–acid battery has been improved. The following data represent the cells for the two batteries being marketed today:

	Edison Battery	Lead–Acid
Open current voltage	1.6	2.2
Typical operating voltage	1.2–1.4	1.95–2.05
Capacity, ampere-hours/lb	10	8
Capacity, watt-hours/lb	13	12

On a volume basis, the lead–acid cell is superior based on both ampere-hours and watt-hours, but the relative capacities of the two batteries will vary considerably depending on the test conditions.

The basic market which Edison sought for the use of his storage battery was self-propelled vehicles, including automobiles, trucks, and streetcars. For such applications he considered resistance to shock and vibration of paramount importance. To say that he tested the battery extensively for shock and vibration resistance would be an understatement. An apparatus was set up in the laboratory which jolted battery cells up an down to simulate heavy road usage. Some were submitted to as many as 1,700,000 such cycles. Automobile tests involving up to 5,000 miles for a single battery were carried out on the roughest roads in New Jersey. Batteries were put in old vehicles and deliberately driven against a stone wall.

Tom Robins, with whom Edison had collaborated in developing conveyor belting,* dropped in one day in 1904 to pay his old friend a visit. As the two men sat talking there was a loud crash just outside the window, as of some heavy object that had fallen; then another, and after a minute, another. Robins was alarmed, but Edison never turned a hair. After a while a workman came in and said, "Second floor O.K., Mr. Edison," Edison nodded and ordered, "Now try the third floor." To the wondering visitor he gravely explained that he was testing the endurance of his storage battery by having packages of them thrown out of the windows of the upper floors of his laboratory. "For a scientist, Edison used some mighty peculiar methods," Robins reflected (9).

The Edison battery was the first alkaline battery developed and marketed. Unlike the acid battery the electrolyte could be housed in a metal container. This, plus the fact that the active

---

* *See* **Chapter 5.**

ingredients were enclosed in containers securely fastened in the grids, made the battery highly resistant to mechanical abuse. Likewise, it was durable when subjected to the chemical abuse of overcharging, overdischarging, or reverse charging. This was likely the result of the fact that the electrolyte undergoes no net change in either charging or discharging.

The advantages of the Edison battery include:

(a) It is the most rugged of all storage batteries.

(b) It has a longer life than any storage battery in use today.

(c) It is virtually electrically foolproof.

(d) It can be stored indefinitely without deterioration.

(e) It can be operated at considerably higher discharge temperatures than can the lead–acid.

(f) At the time it was developed, it provided more energy per unit weight or volume than did the lead-acid battery.

The disadvantages of the battery are:

(a) Voltage is lower than the lead–acid.

(b) Low temperature performance is poor; the battery is rarely used below 0°F.

(c) It is less efficient than the lead–acid type based on ampere-hours or watt-hours obtained on discharge compared with those used on charging.

(d) Performance at high discharge rates is poor.

(e) Considerable hydrogen and oxygen are liberated during charging and standing idle after charging.

(f) The battery is two to three times as expensive as a lead–acid battery of like capacity.

Unfortunately, the electric automobile had largely succumbed to the gasoline powered vehicle before Edison perfected his battery. In the 1890's it was far from obvious which type of vehicle would predominate in the automotive field. The first electric car was commercially produced in 1894 when Henry Morris and Pedro Salom of Philadelphia began producing the

Electrobat. Later manufacturers were Studebaker of South Bend, Anderson of Detroit, the Baker Runabout in the East, and numerous carriage manufacturers who made electric autos as a sideline.

Early in the 1900's, the Electric Storage Battery Company merged with the Pope Manufacturing Company, a producer of automobiles, to form the Electric Vehicle Company. The latter company planned to produce and operate fleets of electric cabs in U.S. cities. About 2,000 such cabs were built but were quite unsuccessful. The lead–acid battery then available and suitable in weight for such vehicles gave only about 25 miles between chargings. The company became known as the Lead Cab Trust and went bankrupt in 1907.

In 1903 the Ford Motor Company was organized and produced 1,700 cars the first year. At that time it was one of 1,500 companies in the United States engaged in manufacturing horseless carriages. A few years later Ford began to market his Model-N, and advertised it as being so simple that a professional chauffeur was not required to drive it. In 1906 a total of 18,000 such cars were manufactured. In October 1908 he introduced his Model-T at the low price of $850. By 1911 this touring car was reduced to $690, and a runabout of the Model-T design sold at $590. In 1914 Ford introduced the moving assembly line and soon was making more than 50% of all cars sold worldwide.

Although Ford and others were rapidly establishing internal combustion as the power for the automobile of the future rather than steam or electricity, it still took a hardy soul to operate any of the non-electric vehicles. The steamers operated at 600 lbs per sq inch pressure, and rumors—mostly unfounded—circulated about the dangers from explosions. The gasoline autos were started by hand cranking. If the spark from the magneto was not properly retarded prior to cranking, the motor usually backfired, reversing the direction of the cranking motion and

Courtesy Edison National Historic Site,
West Orange, N. J.

Electric auto used in testing the Edison battery

often injuring the cranker. So many arms were broken from cranking Fords, that a certain type of broken arm came to be known as the "Ford fracture."

The introduction of the improved Edison battery, which was being produced in quantity by 1910, gave the electric car manufacturers a shot in the arm. A group of the nickel–iron cells of acceptable weight for use in an auto made possible 50–75 miles between chargings. Edison described a trip to Long Island in an electric vehicle where over 100 miles were travelled without battery change at an average speed of 10 miles per hour. Newspapers reported on a prospective trip by the Edisons, using an electric automobile equipped with the new battery, from West Orange to visit Mrs. Edison's mother in Akron, Ohio.

Anderson's of Detroit featured the Edison battery in their electric runabouts. Edison suggested that horses not be allowed in city limits. Said he, "Between the gasoline and the electric cars there is no need for them. A higher public ideal of health and cleanliness is working toward such banishment very swiftly; and then we shall have decent streets, instead of stables made out of strips of cobblestones bordered by sidewalks."

However, the coup de grace to Edison's ambitions for the electric auto was not administered by the mechanical engineers who were improving gasoline-powered vehicles but rather by a fellow electrical engineer. In 1904, fresh from Ohio State University with a degree in electrical engineering, Charles F. Kettering went to work for the National Cash Register Company. He developed the first electrically operated cash register. In 1911, then a partner in operating the Dayton Engineering Laboratories, Kettering used the same principle in developing an electric starter for gasoline automobiles. The new self-starter was successfully demonstrated in Cadillac cars in 1912, and by 1915 it was standard or optional equipment in all American internal combustion vehicles. It was now more the vogue for the fashionable lady to operate a Cadillac with an electrically powered starter than an electrically powered town car. Since a minimum of 6 volts were necessary to operate Kettering's electric starter, the lead–acid battery was the most practical one for starting and lighting the gasoline powered vehicles now surging from the automotive industry. However, up to 1924 the Edison Storage Battery continued to issue data for use of their nickel–iron battery in automobiles and heavy trucks (*8*):

For electrically driven pleasure cars, depending on the size of the battery, they are designed to run fifty to seventy-five miles on a charge, at a speed of about twenty miles per hour.

A five-ton truck will average about thirty-five miles per charge; a three-and-a-half-ton, forty miles; two-ton, forty-five miles; and a one-ton, fifty miles.

Electric trucks were used on city streets longer than were electric automobiles. Department stores found them highly suitable for city deliveries, and many continued to use electric trucks well into the 1920's. The nickel–iron battery continues to find a major use for powering small trucks used to handle baggage and freight at airports, ship docks, railway stations, and bus depots, as well as to move materials in factories.

Edison tried hard to adapt the storage battery to streetcar propulsion. He believed that cost of the electrical energy could be low since the batteries would be charged at central electric stations when some of the dynamos would otherwise be idle in hours of minimum load. Streets would be freed from the burden of overhead trolley wires. Edison built a short railway line on his West Orange property in the winter of 1909–10 and tested a special type of streetcar equipped with motor, storage battery, and other necessary operating devices. This car was later run for a while on Main Street in West Orange and then in New York City. In cooperation with Ralph H. Beach, Edison formed The Federal Storage Battery Car Company with a factory in the Silver Lake area. Such cars found use in isolated areas such as beaches where there was no electric power. However, apparently the economics were unfavorable, and only a few such streetcars were manufactured.

Edison realized that a big disadvantage of his battery was its relatively high cost. The cost of nickel was, and is, a major factor in the cost differential between it and lead batteries. Edison sought new supplies of nickel as indicated by the following quotation from the 1917 report of the Royal Ontario Nickel Commission (10):

Thomas A. Edison, the famous inventor, proposed to enter the field of nickel mining in order to secure a supply of nickel for a newly invented storage battery. He spent a good deal of time and money in diamond drilling on sites located by magnetic instruments, but did not find any deposits of ore.

Although the nickel–iron alkaline battery did not find a permanent outlet in self-propelled vehicles, it continues to have major industrial uses. Besides powering small freight trucks, the battery is used in forklift trucks, mine and switch locomotives, railway signaling, communications equipment, and standby power applications.

The Edison Storage Battery Company was an affiliate of Thomas A. Edison, Incorporated. The main manufacturing facilities for the alkaline storage battery were at West Orange while the chemicals for the battery, as well as the Edison Primary Batteries, were made at Silver Lake. Thomas A. Edison, Incorporated was merged with the McGraw Electric Company in 1957 to form the McGraw-Edison Company. In 1960 the storage battery division was sold to the Electric Storage Battery Company, manufacturer of the Exide lead–acid battery. The nickel–iron battery is now manufactured by the Exide Power Systems Division of that company, now named ESB, Incorporated. Manufacturing was continued at West Orange and Silver Lake until November 1965, at which time all operations dealing with the alkaline battery were transferred to ESB's new plant at Sumter, South Carolina. Only alkaline storage batteries and their electrochemical components are made at Sumter; the plant has 350–400 employees.

Until 1962 the nickel–iron battery was made exactly as developed by Edison. Today the combined potassium hydroxide–lithium hydroxide electrolyte is identical to the original. The positive electrode grids are exactly as developed by Edison except in some cells the perforated tubes are welded in the grid. The 0.00004-inch thick nickel flakes are still used although prepared by a different method. Machinery of the original design continues to be used for filling the positive tubes with alternate layers of the nickel hydrate and metal flake. The negative plate

225

Courtesy Alkaline Battery Division, ESB, Inc.

ESB's plant at Sumter, South Carolina where the Edison battery is manufactured

is no longer a grid of pocket containers, but rather the iron powder is housed in perforated strip-pocket plates. Mercury is no longer used with the finely divided iron. Polypropylene and polystyrene have replaced hard rubber for insulation.

Whereas lead–acid cells last as long as 15–20 years in standby service, the nickel–iron battery has lasted up to 45 years. In extremely deep-discharge service, the nickel–iron battery will last two to three times as long as the lead–acid. The nickel–iron battery of a size comparable with that used in a large automobile sells for about $60. The main reason for the relatively high price is the cost of nickel which is about $1.30 per lb; the price has tripled since 1946. ESB finds that the biggest uses for their alkaline battery are for industrial trucks and standby power.

Sigmund Bergmann, Edison's former employee and manufacturer of electrical equipment for Edison up to the time of the General Electric merger, had returned to Germany and had a thriving electrical business there by the turn of the century. In 1907 he began to manufacture a series of cell sizes of the nickel–iron battery. Once the tubular form of the positive electrode units was developed, the cells found wide use in Germany for railroad motive power and lighting. Some very large cells (4,350 ampere-hours) were built for experimental work by the Italian Navy. The Edison battery has not found use in submarines because of hydrogen evolution during idle standby after charging. The nickel–iron battery continues to be manufactured in Germany in volume and is exported in quantity to other European countries, the Near East, and Africa.

Waldemar Jungner carried out research and development on an alkaline battery about the same time as did Edison. Jungner standardized on nickel and cadmium for electrodes. He claimed that by charging the positive pockets with black nickelic hydroxide, $Ni(OH)_3$, much less swelling resulted with use *vs.* charging with green nickelous hydroxide, $Ni(OH)_2$, as did Edison. The cadmium cycles from the metallic form to the divalent hydroxide.

The main advantage of the nickel–cadmium is that very little hydrogen and oxygen are evolved during charging or discharging, allowing the cells to be hermetically sealed. In spite of the high cost of cadmium, presently $2.50 per lb, the Ni–Cd battery finds commercial use primarily because of its superior high rate-discharge performance and very low self-discharge. The latter property makes it highly attractive for power standby purposes. The production of alkaline storage batteries by ESB Incorporated is about 80% the Ni–Fe type and 20% the Ni–Cd.

The newest type of alkaline storage battery is the silver–zinc. This system offers the highest energy density—*i.e.*, the most watt-hours per unit weight or volume. In addition, it has the least drop in voltage throughout discharge. It also has the lowest self-discharge of any commercial storage cell. This high cost battery finds its main uses in defense applications and the space program.

It is interesting to speculate on the possible course of the automobile in the United States, and the world, if Edison had developed his battery 10 years earlier. The electric auto was never cheap even though it was simpler in construction than the gasoline auto. If the electric car had been the first mass produced automobile rather than the Model-T Ford, it might well have been so cheap that it would never have been completely displaced. The development of the gasoline-powered vehicle was inevitable for long distance transportation, but for city driving it would have been more difficult to replace the electric vehicle with an Edison battery than that with the lead–acid battery available at that time. As every manufacturer knows, once a product is firmly established, it is very difficult to replace it with a more costly although superior product. If the electric automobile had been manufactured in volume, the sales of the Edison battery would have justified a continued major research effort to improve the battery. Standard electrode tests show that the positive electrode requires 4.64 times as much nickel hydroxide as theoretically required to produce one ampere-hour (3). The corresponding value for the negative electrode is 5.78. Thus, theoretically, there is considerable room for improvements in overall battery efficiency, but it would probably take a major research effort to improve the nickel–iron battery over the present efficiency obtained by Edison and the present manufacturer.

The fact that gasoline automobiles would be a source of serious pollution in congested cities was recognized early. Pedro Salom, an electrical engineer, wrote in the *Journal of the Frank-*

lin Institute in 1896 that electric motors had no odor at all, whereas "All the gasoline motors which we have seen belch forth from their exhaust pipe a continuous stream of partially unconsumed hydrocarbons in the form of a thick smoke with a highly noxious odor. Imagine thousands of such vehicles on the streets, each offering up its column of smell!" If the gasoline buggies had replaced fleets of noiseless non-polluting electric cars in city streets rather than the waste-producing horses which attracted swarms of flies, four-legged vermin, and sparrows, the replacement might well have been considered a curse rather than a blessing.

Edison apparently felt that each type of powered vehicle would find its place in American economy as evidenced by the following remarks which he made to Henry Ford when the latter, as an employee of an Edison Illuminating Company, asked his opinion on the gasoline engine for automobiles (*11*):

Yes, there is a big future for any lightweight engine that can develop a high horsepower and be self-contained. No one kind of motive power is ever going to do all the work of the country. We do not know what electricity can do, but I take for granted that it cannot do everything. Keep on with your engine. If you can get what you are after, I can see a great future.

It was a strange coincidence that the man who did the most to make the gasoline automobile such an outstanding success, causing the demise of electric cars and trucks, received heart-warming encouragement from Edison, the man who spent nearly a decade trying to make electrically powered vehicles practical. Also, a company founded on Edison's expertise and patents provided Ford with his livelihood, early engineering experience, and facilities for carrying out his early experiments on the automobile.

Ford obtained a job with the Edison Illuminating Company of Detroit (in 1903 name changed to Detroit Edison Company) in September 1891 as night engineer. He worked a 12-hour shift, starting at 6 p.m.; his salary was $45 per month. He was able to get the job because of his experience with steam engines, and Ford wanted the job to learn more about electricity. He was soon on the day shift, and in November 1893 he was made chief engineer at a salary of $100 per month. By 1896 he was making $1,900 annually and had reached such status in the company that he was one of the representatives of the Detroit company attending annual meetings of the Association of Edison Illuminating Companies.

As soon as Ford was well established at the Edison Company he proceeded to use the shops there while working on his gasoline powered vehicle (12). If Ford had had a contract with the Edison company of the type which is accepted practice today between technical employees and their employers, the early developments of Ford's experimental automobile would have belonged to the Edison Illuminating Company. As it turned out, the officials of the Edison company thought Ford a misled young man and, although appreciating his mechanical abilities, felt that his tinkering with his automobile detracted from his progress in the more promising field of electricity. In 1899 Alex Dow, President of Edison Illuminating Company of Detroit, offered Ford the job of plant superintendent, pointing out at the time that such a responsible job would not allow him any time for his automotive interests. Ford gave no thought to giving up the latter since he felt that the future of the automobile was then assured. He resigned from the Edison Company in August 1899.

Edison and Ford soon became great friends even though they were working in opposite directions in the early developments of the automobile. In his visits East, Ford would usually drop in at the West Orange Laboratories. Edison advised Ford that gears should be preferable to chain drive for his automobile. Later he advised against trying to adapt the generator, used for charging the storage battery, to the additional role of electric motor for the starter. Ford credited Edison's automation at his ore concentration plant at Ogdensburg with sparking the idea for his assembly line operations at his Highland Park plant and later at River Rouge. Ford's blind belief that Edison could do no wrong led him to install direct current in his River Rouge plant although it was rather obvious at the time that alternating current was the electricity of the future. This resulted in an expensive changeover from direct to alternating electric motors beginning in 1930.

Ford described Edison as "The world's greatest scientist. His knowledge is almost universal. He believes all things are possible. At the same time he keeps his feet on the ground." Ford was also known to comment that Edison "might well be the world's worst business man."

Following Ford's announcement in January 1914 of his minimum $5-day wage and installation of three shifts a day instead of two, Edison, in an interview with the *New York World,* praised such "radical innovation." He pointed out that Ford's technique of using special machinery and highly efficient methods were open to every line of business. When Ford was considering running for the President of the United States after losing a close contest for U.S. Senator in 1918, Edison said of Ford, "He is a remarkable man in one sense, and in another he is not. I would not vote for him for President, but as a director of manufacturing or industrial enterprises, I'd vote for him— twice." Edison later used "Friend Ford's wonderful touring car" for his travelling both at West Orange and Fort Myers, preferring it to the larger cars used by his wife.

Most major developments in applied science result when the time is ripe. That is, it is first necessary to have the necessary basic knowledge, and suitable raw materials and process equipment must be available. If Jones doesn't make a rather obvious invention, Smith will do it in a few months or years. As is well known, two or more investigators often make the same invention independently at nearly the same time. Alexander Bell beat Elisha Gray to the Patent Office with the telephone invention by only a few hours. Synthetic rubber was not possible until it was shown that natural rubber is a combination of isoprene units. The practical electric light could not have been developed before the dynamo. The automobile would have been impossible without vulcanized rubber of high abrasion resistance. Synthetic diamonds were not possible until we had materials which would stand very high pressures at high temperatures—and so on.

Edison's development of the nickel–iron alkaline storage battery is one of relatively few inventions which do not fall into this one-thing-leads-to-another category. Furthermore, if Edison had been working for a large corporation, it is very likely that his positive electrode would have been considered a "Rube Goldberg device" and would have been the subject of many jokes but no sales and profits. The following is a partial list of the bad features of such an electrode which the experts responsible for new product development would have likely pointed out:

(a) Both the oxidized and reduced forms of the nickel are non-conductors and thus unsuitable for use as an electrode.

(b) The nickel flake is not only very expensive to prepare, but is so fluffy it is next to impossible to handle.

(c) Because of the non-conductivity of the nickel hydroxides, they are housed in very small diameter metal tubes. This complicates their filling and the need for 30 such tubes in order to construct a single grid.

(d) There are 600 separate charges of material to each 0.2-cubic inch tube plus 200 separate stampings of 2,000 lbs/sq inch each during the filling operation.

(e) In compressing the contents of the tubes at such a high pressure in order to get conductivity, the nickel hydrate becomes so dense that the electrolyte only partially penetrates the tube contents. This results in the necessity of using 4.64 times the theoretical amount of nickel hydroxide.

(f) Swelling of the tube contents during charging makes necessary the use of reinforced rings around the tubes to avoid their bursting. This adds to the cost and weight of the cell.

Thus it is considered very unlikely that the nickel–iron battery would have ever been manufactured and marketed if someone as strong-minded as Edison had not worked on it and was in a position to go into production without getting the approval of several levels of management. These facts plus the additional fact that the nickel–iron battery as currently manufactured is substantially as Edison developed it over 60 years ago, the fact that the battery was largely the development of one man, and the fact that it is still an important article of commerce make the nickel–iron alkaline storage battery one of the outstanding developments in the history of applied chemistry.

Chapter 8 Organic Chemicals and Naval Research

THE SPANISH–AMERICAN WAR of 1898 was primarily a naval war, and Edison foresaw that technology could be an important factor in such conflicts. In 1898 he suggested to the Navy Department the use of a shell containing a mixture of calcium carbide and calcium phosphide for making enemy ships visible at night. When such a shell was exploded over water, the carbide would react on contact with water and liberate acetylene; likewise, the phosphide would liberate phosphine. The latter, PH_3, ignites spontaneously on contact with air because of the presence of an impurity, P_2H_4. Burning of the acetylene would add volume to the fire. Such a flare would continue for several minutes as the solid calcium carbide particles react with water.

Edison also aided Winfield Scott Sims in developing the Sims-Edison torpedo. Sims, an American inventor, had served in the Civil War. The Sims-Edison torpedo was hung from a float in such a way as to be held a few feet below the surface of the water. It was equipped with a screw propeller and rudder. It contained, in addition to the explosive charge, a small electric motor that furnished driving and steering power. When fired, it trailed behind it an electric cable through which it could be controlled. The torpedo was found to be so slow that its practical value was seriously impaired, and before long it became obsolete. However, by using batteries for the source of electricity rather than a trailing cable this torpedo design pioneered the changeover from compressed-air engines to electric motors.

Edison foresaw the increasing horrors of war, and shortly after the Spanish-American conflict he caused considerable consternation in Europe and America by the following prophecies (*1*):

Electricity will play an important part in the wars of the future. Torpedo boats can be despatched two miles ahead of a man-of-war and kept at that distance under absolute control, ready to blow up anything within reach. I believe, too, that aerial torpedo boats will fly over the enemy's ships and drop a hundred tons of dynamite down on them. A five-million-dollar war vessel can be destroyed instantly by one of these torpedoes.

I can also conceive of dynamite guns. I have no intention of ever devising machines for annihilation, but I know what can be done with them. Nitroglycerin is one of the most dangerous substances that man can deal with. Touch a drop of it with a hammer and you will blow yourself into the hereafter. Iodide of nitrogen is even more dangerous. While experimenting with explosives in magnetic mining, I made some of them so sensitive that they would go off if shouted at. Place a drop on the table and yell at it and it will explode.

It is interesting that these predictions were made before the first powered air flight—Wright brothers, December 17, 1903—or the development of Zeppelin lighter-than-air craft. Edison's "aerial torpedo boats," which in reality became torpedo planes, certainly sank a number of warships in World War II. High explosives replaced low explosives of the black powder type by the time of World War I when the manufacture of TNT was limited only by the availability of toluene.

Starting about 1900 Edison's research and development studies became more chemical in nature. As noted in Chapter 4, he and his co-workers pioneered work on the preparation and molding of various plastics compositions. His work on the manufacture of phenolic resins deserves a more complete story.

As a matter of fact, Edison's use of phenolic resins for phonograph records was one reason why he went into the manufacture of organic chemicals during World War I.

Early work on phenolic resins is usually reported in the chemical literature to be largely that of one man, Leo Hendrik Baekeland. A Belgian chemist, Baekeland migrated to the United States in 1889 and founded Nepera Chemical Company. He developed a photographic paper, which he called Velox, that could be developed in artificial light. He sold this development to George Eastman in 1899 and as a result became independently wealthy. Finding retirement not to his liking, he began work on the phenol–formaldehyde reaction in 1905 at his home in Yonkers, New York.

In 1872 the famous German chemist, Adolf Baeyer, reported that in the presence of mineral acids phenol and formaldehyde condense to give a variety of products. In 1894 other workers showed that alkalies were also effective for promoting this condensation, and at low temperatures the reaction could be controlled to form 2-methylolphenol from one molecule each of the reactants. Arthur Smith was granted the first patent on phenolic resin in 1899, which he described as a substitute for hard rubber (2). This and other attempts at this time to commercialize the phenolics failed.

Baekeland undertook the study of the phenol–aldehyde reaction with the primary objective of finding a substitute for shellac which at that time was being imported into the United States at the rate of about 50 million lbs annually. Early in his work he determined that simple phenol and formaldehyde were the best of the two classes of compounds to employ. He studied both alkaline and acid catalysts and various process conditions to get reproducible results. His two main contributions were:

(a) Use of a filler, such as wood flour, to overcome the brittleness of the resin.

(b) Use of heat and pressure during molding.

The latter development led to Baekeland's famous "Heat and Pressure Patent," U.S. 942,699, issued December 7, 1909. This patent was the bulwark of Baekeland's patent position. It is surprising that such a patent was granted and sustained. Heat and pressure had been used for years to cure (vulcanize) rubber, including hard rubber which is a thermoset plastic as is cured phenolic.

When alkali is used as catalyst for the phenol–formaldehyde reaction, the first product is methylolphenol, which in turn condenses with itself:

$$\text{OH} \quad + \quad HC\,OH \quad \longrightarrow \quad \text{OH} - CH_2OH$$

$$2 \;\; \text{OH} - CH_2OH \quad \longrightarrow \quad \text{OH} - CH_2 - \text{OH} - CH_2OH + H_2O$$

This reaction is usually with 37–40% aqueous formaldehyde with enough sodium hydroxide added to maintain a pH of 7–11. If the temperature is controlled, the degree of condensation can be kept below the gel point. Such solutions can be used to impregnate paper or cloth to make laminates, or the resin can be recovered as such by evaporation of the water. If one or more moles of formaldehyde are used per mole of phenol, the resin can be set to a non-fusible state simply by heating. Since the para position of the phenol is reactive in the condensation reaction as well as both ortho positions, highly crosslinked molecules can be formed.

When aqueous formaldehyde and pure phenol are mixed, the solution has a pH of 3.0–3.1. This acidic pH is the result of the fact that phenol is a weak acid. If the pH is lowered to 0.5–1.5 by adding a mineral acid, rapid condensation occurs. If the molar ratio of aldehyde to phenol is one or higher, uncontrollable setting of the resin results. If, on the other hand, a molar excess of phenol is used, stable fusible and soluble condensates known as novolak resins result. With acid catalysis, the methylolphenol is more reactive with phenol than it is with a second molecule of methylolphenol—*i.e.*:

$$\text{(phenol-CH}_2\text{OH)} + \text{(phenol)} \longrightarrow \text{(phenol-CH}_2\text{-phenol)} + \text{H}_2\text{O}$$

Unlike the one-step resins, the novolak resin is a completed reaction, and even if heated at the usual cure temperature range of 350°–400°F, it will not thermally set.

Jonas W. Aylsworth in collaboration with Edison was also working on phenolic resins during the latter part of the first decade of the 1900's. It is not clear whether Aylsworth knew of Baekeland's promising results before he initiated his studies on the phenolics. As previously mentioned, others had worked on phenol–formaldehyde resins prior to Baekeland, and concurrently Hans Lebach of Germany had been studying the best methods for molding phenolics. On February 5, 1909 Baekeland discussed his new Bakelite resins before the New York Section of the American Chemical Society (3). The cured resins were described as being more scratch-resistant than shellac and harder than hard rubber. One of the uses suggested was for phonograph records. In another paper presented before the same group on May 4, 1909, Baekeland discussed the novolak resins, which was at that time a trademark name (4).

Aylsworth's initial objective was to use the phenolic resin as a varnish for coating phonograph blank records. Since this would be a casting and would be set with little pressure, it was essential to avoid the evolution of water vapor which is, of course, a by-product in the setting of such alkali catalyzed resins. He accomplished his objective by adding hexamethylenetetramine (hexa) to the novolak-type phenolic and heating. Hexa, $(CH_2)_6N_4$, is a condensed heterocyclic compound having three six-membered rings and is prepared by condensing equal molar proportions of ammonia and formaldehyde. Subsequent splitting out of water followed by polymerization and elimination of a part of the combined ammonia results in the hexa. Since hexa is an amine, it converts the novolak from an acid medium to alkaline. When mixed with the novolak resin and heated at 200°F and higher temperatures, the hexa functions as a crosslinking agent which converts the low melting phenolic resin to a higher melting one and finally to an infusible thermoset plastic.

Aylsworth filed a patent on February 11, 1910 covering the use of hexa and allied materials for curing phenolic resins (5). This application issued first as a Belgian patent. In a "Letter to the Editor" in July 1911 (6), Baekeland commented on the Aylsworth patent to the effect that nothing new was disclosed since he had reported ammonia to be a catalyst for the formaldehyde condensation with phenol or novolak resins. Baekeland maintained that hexa was equivalent to formaldehyde plus ammonia. His comments were scientifically unsound in that although hexa is prepared from formaldehyde and ammonia, it does not decompose to formaldehyde and ammonia when heated. The oxygen from the formaldehyde is completely eliminated in the synthesis of hexa. Although hexa is capable of condensing with low molecular weight phenolic resin to provide additional methylene linkages as does formaldehyde, it does so without forming water as a by-product.

The Condensite Company of America was organized September 23, 1910, to manufacture phenolic resins under the

Aylsworth patents. Initial manufacturing operations were at Edison's battery plant in Glen Ridge. Kirk Brown, a friend of Frank Dyer, head of Edison's legal division, helped to finance the new enterprise and became president. A new plant was built in neighboring Bloomfield and occupied in 1914. A loan was obtained from the Bank of Montclair to help finance the new factory, with Dyer and others, as well as Kirk, having an interest in the company.

The General Bakelite Company was formed September 29, 1910 following small scale pilot-plant operations at Baekeland's laboratory on his estate in Yonkers. Baekeland had a controlling interest in the new company, with initial operations in a building leased from Roessler and Hasslacher Chemical Company at their Perth Amboy, New Jersey plant. This company supplied the formaldehyde for the new operations and assumed most of the management and operation of Bakelite. In 1913 the General Bakelite Company built a three-story building on land which had been purchased adjacent to the R and H Chemical plant in Perth Amboy.

In 1913 Bakelite sued Condensite for patent infringement. This initial suit was withdrawn with Condensite agreeing to pay royalties for use of certain Baekeland patents. However, practical arrangements did not work out, and Bakelite brought a second suit, which in 1917 was decided in favor of Condensite (7). By this time Aylsworth had established an imposing list of patents on phenolics; he had obtained a total of 21 in 1914 alone. The record in the chemical literature is confusing as to just what constituted the licensing agreement worked out between Bakelite and Condensite. The statement by Haynes (7) is probably not the whole story. Apparently Condensite continued to recognize the validity of Bakelite's "Heat and Pressure" patent, U.S. 942,699 and that dealing with wood flour and other fillers, 942,852. On the other hand, Bakelite apparently recognized the validity of several of Condensite's patents, including those covering dry compounds containing hexa.

Laurence V. Redman, a graduate of Toronto University, was the third figure to enter the phenolic arena. In 1910 he accepted an industrial chemical fellowship at the University of Kansas and became an assistant professor of industrial chemistry there. The fellowship was established by Adolph Karpen of Chicago to study new varnishes for furniture. Redman's studies concentrated on phenolics prepared primarily by the reaction between anhydrous phenol and hexamethylenetetramine. Thus, his work was similar to Aylsworth's but different in that Redman allowed hexa to react with phenol, whereas Aylsworth first prepared the fusible phenolic from phenol and formaldehyde and then cured the anhydrous resin so obtained with hexa. Redman apparently had worked on phenolics before coming to Kansas since he filed his first patent on the subject June 17, 1910.

Early in 1914 Redman and two of his associates at the Department of Industrial Research, University of Kansas, published a paper dealing with synthetic resins in the American Chemical Society journal *Industrial Engineering Chemistry* (*8*). In this paper the authors gave a good discussion of the chemistry of the phenolic resins and listed numerous references, including 15 patents obtained by Baekeland. They also discussed their own work briefly.

Baekeland promptly verbally blasted the Redman *et al.* paper (*9*) even more caustically than he had the patent of Aylsworth (*6*). He commented that, "The paper in question would have had increased importance if some of the opinions and statements expressed therein were more in accordance with facts." Baekeland also criticized the University of Kansas for allowing its facilities to be used "in the interests of purely commercial enterprises." Redman and his co-authors replied (*10*) that their work was quite different from that of Baekeland in that they used a dry system. In a critical mood themselves, they commented that Baekeland was trying to use his Bakelite trademark on a much broader class of phenolic resins than was justified. They maintained that Baekeland's contribution was "to study the work of others and to put to industrial uses the results of their labors supplemented by his own discoveries."

Redman and his associates at Kansas, financially backed by the Karpen family, in 1914 founded the Amberoid Chemical Products Company. Initial operations were in a Karpen furniture factory on West 22 Street in Chicago. This company began to manufacture a phenolic resin under the trademark name of Redmanol. The name of the company was shortly changed to Redmanol Chemical Products Company.

During 1916 Baekeland and Redman continued to exchange comments on the subject of who contributed what to the development of phenolic resins in several "Letters to the Editor" of *Industrial Engineering Chemistry*. Baekeland compared his 34 patents on the subject with only one obtained by Redman. Redman contended that a patent to Luft in 1903 anticipated Baekeland's work with alkaline catalysts. Late in 1916 Kirk Brown of the Condensite Company of America also got into the act. After Aylsworth died in June 1916, Brown apparently felt obligated to support Aylsworth's position, even though Aylsworth, as Edison, had shunned the printed page. Excerpts from Brown's "Letter to the Editor" are given below since his comments were a more rational evaluation of the situation than those from the other two parties (*11*):

In the May and June, 1916, numbers of This Journal, there is an interchange of letters between Dr. L. V. Redman and associates and Dr. L. H. Baekeland, bearing on the art of phenol condensation products. We have no desire to take sides in that controversy, but as Dr. Redman in initiating his correspondence gives as his reason the necessity "to preserve an accurate record of the scientific and industrial development of synthetic phenol resins," mention should be made of the achievements of the late

J. W. Aylsworth to whom was granted, for inventions in this field, some 47 United States and many foreign patents, more than those of any other inventor.

Dr. Redman reviewed the work of Dr. Baekeland with relation to that of those who preceded him, pointing out in many particulars how Dr. Baekeland's work differed sometimes but little, or again unimportantly from that of others. Without going into the merits of these contentions, we would like to record our appreciation of the work that Dr. Baekeland has done as a whole in the field of phenolic condensation products. Whatever the numerous earlier inventors who have worked on the phenol-methylene reaction did, there was certainly one thing that all of them together did not do, and that was to reduce any of their inventions to a commercial possibility. Dr. Baekeland was the first to do this, and it does not detract from the value of his work that it should have been done almost simultaneously with the like advent of the work of J. W. Aylsworth.

The Aylsworth process involving the use of hexa is, as stated, earlier than those of the others, is broadly protected, and, we believe, has important advantages. According to the main Aylsworth process, a phenol resin is first prepared, in which the proportion of methylene is such that the resin is permanently fusible (unchanged when heated to 420°F or higher). Therefore it can be heated to a high temperature for complete dehydration. It was found by Mr. Aylsworth that about 400°F of heat is required to remove all the water, free and combined. Such dehydration of a partial condensation product, of the type which becomes infusible on further heating, would of course be impossible, because the mass would be hardened long before the desired temperature was attained.

To form the final fusible product, this dehydrated, fusible resin is combined with the necessary percentage of hexa (or with paraform, Aylsworth patent No. 1,102,630). One advantage of using hexa is that the reaction is anhydrous, ammonia being evolved but no water, formaldehyde or other gaseous products. No other counteracting pressure is needed.

After the suit between Bakelite and Condensite, Bakelite agreed to sue any third party who infringed the patents of

243

either company. Subsequently Bakelite sued General Insulate Company, a customer of the Redmanol Company. Redmanol defended the suit. Years passed, but on August 2, 1921, a court decision was rendered against General Insulate and thus against Redmanol. However, Adolph Karpen, anticipating a possible adverse decision, had secretly bought the financial interests of the Bank of Montclair, Frank Dyer, and others in the Condensite Company. This gave Karpen a controlling interest in Condensite, and he confronted Baekeland and Kirk Brown with the fact that they had a bedfellow whether they wanted him or not.

The upshot was that in March 1922 the three companies merged to form the Bakelite Corporation as a holding company, with each company continuing to operate under its respective trademarks. Baekeland became president, Kirk Brown, vice-president and general manager, and Redman and Adolph Karpen, vice-presidents. Redman was in charge of research and development. In 1924 the Corporation became an operating company with manufacturing coordinated at the three sites. Starting in 1929 a new plant was built at Bound Brook, New Jersey. When this operation came into full production in 1932, Baekeland's original plant at Perth Amboy and that of Redman's at Chicago were shut down. However, the Condensite plant, which in 1925 had become the site of all research and development for the Corporation, continued operations until 1958. Union Carbide Corporation acquired the Bakelite Corporation in 1939, at which time Baekeland retired.

Both Baekeland and Redman had illustrious chemical careers, each winning several scientific awards. Baekeland was elected President of the American Chemical Society in 1924, and Redman held the same office in 1932. Following the merger the two became good friends, the older and more experienced Baekeland introducing his protege to the politics of the chemical profession. In 1928 Redman wrote a highly complimentary sketch of his boss (*12*). Both were excellent speakers and definitely the joiner type. Their combined efforts were instrumental in establishing phenolics in the field of molding compounds, as adhesives in the manufacture of laminates, plywood, and veneers, as bonding agents for abrasives and foundry molds, and for coatings. The phenolic industry literally ushered in the plastics age, and it remains today the largest volume of all thermosetting plastics. After the basic patents held by the Corporation expired in the late 1920's, others entered the field, and today there are over 50 manufacturers of phenolic resins in the United States alone.

By 1914 the United States was a world leader in electrochemistry, and its inorganic or mineral chemical industry was comparable with that of Germany or any other country. In plastics the United States led the world; both hard rubber and celluloid originated here. The new phenolic resin industry had its roots in the United States. Because of the booming automobile industry, we led the world in rubber consumption. However, in organic chemistry we were far behind the European powers. At that time, the primary source of raw materials for the manufacture of organic chemicals was the volatile portion from the carbonization of coal to make coke. Prior to World War I, the United States imported more than 65% of its coke. Furthermore, the domestic coke was produced primarily in beehive coking ovens where the only product was coke; all volatile products were used as fuel for the coking operation or in the manufacture of steel.

Besides gaseous hydrocarbons such as methane, the chief chemicals derived from the coking of bituminous coal are ammonia, benzene, toluene, the xylenes, phenol, naphthalene, tar bases such as pyridine, tar acids, and pitch. If the so-called coal tar chemicals are to be recovered, a coking oven of special design is used, equipped with apparatus that condenses or otherwise recovers the various volatile products. The hot gases and vapors which leave the furnace at over 1,000°F are first sprayed with a recycle liquor to remove ammonia and cool them prior to condensation and isolation of the liquid products. The first coke ovens in the United States designed to recover by-products were constructed by the Semet-Solvay Company at Syracuse, New York in 1893. This company's principal interest was the recovery of ammonia to use in the manufacture of soda ash.

In 1910 Semet-Solvay joined with Barrett Chemical and General Chemical to form the Benzol Products Company. Solvay supplied the crude raw material from coke ovens, and the new company separated it into creosote oil and pitch for wood preserving, a crude benzene fraction for enriching illuminating gas, and a small production of high grade benzene and toluene. About 0.7% of phenol was in the crude by-products from the coker, and 0.25% was recovered by extracting with aqueous sodium hydroxide. In 1913 there were about 16 other small coking operations which recovered some by-products. Their primary product besides pitch and tar products was a so-called light oil which was used to enrich manufactured gas.

When World War I began in August 1914, the Allies imposed a naval blockade on Germany. They also placed embargoes on the export of most of their own chemicals, because of their strategic value in the war. Because of the General Bakelite's close relationship with Roessler & Hasslacher, a company of German origin, they apparently had sufficient foresight to have a good supply of phenol on hand. The Condensite Company was caught with very little inventory. It is obvious that at this time Condensite was operated and largely under the control of Edison since it was he—not the Condensite Company—who took steps to make synthetic phenol. In the fall of 1914, Edison required about one ton of phenolic resin per day to manufacture

Edison's laboratories, such as the one shown here, allowed him to move rapidly into the manufacture of synthetic phenol and other organic chemicals.

phonograph records. He was by far the largest consumer of phenolics in the United States at that time. Bakelite's production in 1914 was 783,000 lbs, an average of about 1.1 tons per day. This, of course, went to a variety of customers. Condensite also sold to others, but Edison was its principal customer.

The Semet-Solvay Company had prepared synthetic phenol in 1900 on a pilot-plant basis for the primary purpose of furnishing experimental quantities of picric acid to the U.S. Army. Picric acid, trinitrophenol, is easily prepared by nitrating phenol. The French began using this explosive in large

247

shells in 1885. Semet-Solvay did not go into phenol manufacture since at that time the market for this chemical was only about a million pounds a year, primarily as a disinfectant.

At the start of hostilities in 1914, the supply of phenol not only decreased because of cessation of imports, but shipment of picric acid to the Allies was limited only by supplies. Edison approached several chemical companies with an offer of a long term contract to purchase phenol, with the hope of encouraging one or more of them to go into synthetic phenol manufacture. None would promise delivery within less than six to nine months. In characteristic fashion, Edison decided to manufacture it himself. The process—sulfonation of benzene followed by caustic fusion—was described in the literature, but production details were largely trade secrets. As every chemical manufacturer knows, it makes an enormous difference whether one is manufacturing by the ounce or by the ton. Edison assigned 40 of his staff to laboratory, design, and procurement studies on a three-shift basis. As details of the process steps were checked out in the laboratory, the equipment was designed. Within a week plant construction began at Silver Lake on a three shift seven-day week basis. Within 18 days the plant began production, and within a month after initial work had started on the project, a ton a day of phenol of better purity than the natural product was being manufactured. A month or so later, the plant achieved a production capacity of six tons a day.

Phenol manufacture by the sulfonation fusion process is not simple. The steps involved are as follows:

(a) Benzene is sulfonated using an excess of 98% or stronger sulfuric acid.

(b) The benzenesulfonic acid and excess sulfuric acid are converted to their calcium salts by addition of an aqueous slurry of calcium hydroxide.

(c) The insoluble calcium sulfate is filtered from the soluble calcium benzenesulfonate.

(d) Sodium carbonate is added, and the precipitated calcium carbonate is separated from the solution of sodium benzenesulfonate.

(e) The solution from (d) is evaporated to dryness in a vacuum oven.

(f) Solid sodium hydroxide is heated to fusion, and the dry sodium sulfonate is added slowly at 300°C.

(g) The melt is run into pans and cooled. It is broken up and then acidified with carbon dioxide or sulfur dioxide to liberate the phenol.

(h) The crude phenol is separated from the aqueous salt layer and purified by distillation.

Considering there are 10 separate operations in this synthesis and that such chemicals as concentrated sulfuric acid, molten caustic soda, and phenol were handled, the building of such a plant within a few weeks was no mean achievement. Edison later remarked, "I take a great deal of pride in the fact that within 60 days after we decided to make carbolic acid, we had built a plant and were doing it." Some of the equipment was intricate; for example, a silver-lined condenser was used in the final distillation to obtain a product of good color. One reason Edison could "throw a plant together" so quickly was because he had shops capable of fabricating factory-size equipment. When he purchased, his prestige tended to give him priority. However, Edison did not build his phenol and other chemical plants for long term, efficient operation. His objective from the start was to supply strategic chemicals only during the war emergency. He apparently felt that on a long term basis he could not compete with the integrated chemical companies but that the inflated prices during the emergency allowed him quick pay-out.

Edison sold the excess production from his six-ton-per-day phenol plant which was not required by Condensite. Demand was so great he soon built a second phenol plant of like capacity, using some improved process steps which his laboratory had developed in the interim period. To provide adequate benzene feedstock for his chemical plants, he contracted for crude liquid fractions obtained in coal tar recovery operations at the Cambria Steel Company, Johnstown, Pennsylvania, and the Woodward Iron Company, Woodward, Alabama. Recovery and purification equipment were erected at both sites, and not only was benzene made available but also toluene, mixed xylenes, solvent naphtha, and flake naphthalene. The Cambria installation recovered 18,000 gallons of benzene a day. Recovery of high quality toluene was particularly significant after the United States entered the war since TNT was the favored explosive for munitions and was used instead of other explosives as the supply of toluene allowed.

Another chemical which Edison needed for his phonograph business was paraphenylenediamine. This was used in catalytic amounts in the blend of phenolic resin and hexa in the surface veneer of phonograph records. Edison himself had a patent on this, and he claimed that its use resulted in a more perfect surface, free from bubbles and other flaws. Paraphenylenediamine is bifunctional and thus is more difficult to prepare than is a monofunctional compound such as phenol. To make this diamine, Edison had to prepare several other chemicals which were in short supply and which he marketed on a limited basis. These included the following:

Nitrobenzene was prepared by treating benzene with a mixture of nitric and sulfuric acids. The nitric acid attacks the benzene preferentially; the sulfuric acid acts as a solvent for the nitric and the by-product water. Nitrobenzene was an important industrial solvent and also found use as an accelerator in vulcanizing rubber with sulfur.

Aniline was prepared by reducing nitrobenzene with iron filings in acetic acid. Edison's notebooks show that it is also possible that iron filings in hot sodium hydroxide may have been used. Edison not only marketed aniline but also aniline hydrochloride. Both were important dye intermediates.

Acetanilide was synthesized by acetylating aniline with acetyl chloride. Acetanilide was in demand as an accelerator for vulcanizing rubber and had use as an antiseptic.

Paranitroaniline was prepared by the nitration of acetanilide followed by hydrolysis of the acetyl group to acetic acid. This nitroaniline was used in the dye industry.

Paraphenylenediamine was prepared by reducing *p*-nitroaniline with iron filings and hydrochloric acid. Use of hydrochloric acid with the iron is necessary in this reduction to convert the amino groups to the hydrochlorides and thus avoid side reactions. The free diamine was liberated from its salt by sodium hydroxide, the iron hydroxide was removed by filtration, and the water-soluble diamine was concentrated and crystallized from the water solution.

Edison first set up a pilot plant at his West Orange Laboratory to make 25 lbs a day of *p*-phenylenediamine which was more than adequate for his requirements. However, this diamine was the only chemical known at that time for dyeing furs; when applied to fur, it turns black on exposure to air. Once the manufacture of the diamine became known, Edison was besieged by fur dyers for supplies. He proceeded to build a commercial plant to make 250 lbs a day of the *p*-phenylenediamine, and production was soon increased to 1,000 lbs. This not only supplied the U.S. market, but some was exported to Europe and Japan. When the commercial plant was built, the production of the intermediates cited above was expanded to supply markets for these chemicals as well as for their use in the manufacture of the *p*-phenylenediamine.

Edison also manufactured *p*-aminophenol. Phenol was nitrated with cold dilute nitric acid, and the ortho isomer was removed by steam distillation. The *p*-nitrophenol was then reduced to the aminophenol. The aminophenol was used in photography and as a medicinal. Another important dye intermediate which Edison manufactured was benzidine. This was a rather intricate synthesis for a "mucker," as he chose to call himself. Nitrobenzene was reduced with zinc dust in alkaline solution with gradual addition of reactants at about 100°C:

$$2 \quad \text{C}_6\text{H}_5\text{NO}_2 \; + \; 5\,\text{Zn} + 10\,\text{NaOH} \longrightarrow$$

$$\text{(Hydrazobenzene)} \; + 5 \; \text{Na}_2\text{ZnO}_2 + 4\,\text{H}_2\text{O}$$

Hydrazobenzene

The solid hydrazobenzene was removed by filtration, washed, and slurried in ethyl alcohol. Acidifying the latter with sulfuric acid at a slightly elevated temperature resulted in rearrangement to benzidine:

$$\text{NH}-\text{NH} \; \xrightarrow{\text{H}^+} \; \text{NH}_2 \quad \text{NH}_2$$

Benzidine was an important dye intermediate, particularly for Congo Red.

Edison brought all of his emergency chemical operations on stream either in late 1914 or in 1915. By 1916 many others had built organic chemical plants. During World War I over a dozen synthetic phenol plants were built in the United States; their capacity reached 160 tons per day. Whereas in late 1914 the price of phenol rose to $1.50 per lb and that of benzene to $1.20 per gal, by mid-1916 phenol was selling for

65¢ and benzene for 60¢. By 1917 phenol was in surplus, and Edison shut down his two plants. By 1919 the price of phenol was 8¢ per lb. The U.S. Government had a stock of 35 million lbs of phenol at the end of the war. At the request of the Government, Dow Chemical Company built a large phenol plant in 1917, and at the end of hostilities in November 1918, Dow had an inventory of 4 million lbs. Their phenol plant was closed down and in 1919 was dismantled. By the end of the war, Edison had substantially ceased to manufacture all chemicals initiated in the 1914–15 period. His plants had not been built for long term operation, and he was wise to close them early while his equipment could be readily disposed of because of wartime shortages.

Soon after the war began in 1914, the United States housewife found that the color of new family clothes was likely to change drastically or even to disappear when laundered. Newspapers raised a hue and cry about the shortsightedness of our chemical industry in not manufacturing dyes and other essential organic chemicals. The failure of the steel companies and others to recover chemicals from their coking operations was branded as near criminal. The attempt to find adequate scapegoats was a rehearsal for the clamor which arose in 1942 when the Far East was shut off as a rubber source and newspapers discovered that we had little in the way of a synthetic rubber industry.

Since Edison was the first to produce many chemicals in short supply, he received a great deal of favorable newspaper publicity although his efforts were largely ignored in the chemical press. In fact, Edison was criticized by A. H. Ney, a consulting engineer, in an article published in a leading chemical journal for allowing newspapers to publish "that the old and well known process (for making phenol) was claimed as his own invention, and the appearance of exaggerated and foolish stories in the newspapers was permitted and countenanced." What

Dr. Ney did not appreciate was that Edison always ignored what the news media and technical press said about him—good or bad. He never claimed to have worked out any new process for any organic chemical, the manufacture of which he initiated during World War I. This is further apparent since no patents were issued to him on any of these operations. However, one has only to look over the laboratory notebooks used during late 1914 through 1915 at West Orange to realize that he did a great deal of laboratory work to optimize the various process steps. That Edison played an active part in this work is evidenced by the fact that one repeatedly finds the entry, "Turned sample over to Mr. Edison for analysis." Apparently the procedure was for Edison to examine individual samples, check their syntheses, and see that they were analyzed to his satisfaction.

Shortly after the European war broke out in 1914, both Edison and his friend Henry Ford became well known for their pacific attitudes. Edison commented that the war made him "sick at heart" and he would have no part of it. "Making things which kill men is against my fiber," he added. He maintained this policy in his production of strategic chemicals. Although he manufactured phenol and several of its derivatives, he did not make picric acid, the explosive derived from phenol.

Edison's policy that his chemicals go into domestic uses gave rise to what was called the "phenol scandal." Early in 1915 when Edison's production of phenol exceeded his requirements, the surplus was marketed by the American Oil and Supply Company, a well-known Newark chemical jobber. A considerable quantity of this surplus ended up at the Heyden Chemical Company, where it was converted to acetylsalicylic acid and sold to the American subsidiary of Bayer and Company for marketing as aspirin. Since both of these chemical companies were German controlled, some newspaper columnists proclaimed that Edison had been duped by Germans who wished to keep the phenol from being used to make picric acid for the

Allies. The truth is that Edison apparently preferred to have his excess phenol converted to aspirin for Americans rather than to an explosive for the European war.

Edison exhibited a broad knowledge of chemical know-how and self-reliance when he began to manufacture a variety of organic chemicals in 1914–15. Although some chemists were employed in his laboratories and plants, it was Edison who made the decisions as to plant design and operation. Actually he didn't think too much of college-trained chemists. For example, he hired a mechanical engineer and assigned him to a project in the chemical laboratory. Highly agitated, the engineer sought Edison to point out that he was not a chemist. Edison quickly assured the new employee that he was well aware of that, saying "If I had put a college-trained chemist on the job, he wouldn't have tried anything new since as a chemist he knew all the answers." Likewise, Edison often used college-trained chemists on non-chemical jobs. Theodore Lehman, who was a chemistry graduate from Columbia University and had obtained a Ph.D. in chemistry in Germany, was put in charge of Edison's iron ore concentration plant in Norway.

Edison did not hire any big-name organic chemical consultants to advise him in his 1914–15 venture. He did hire a consulting engineer, Victor L. King, to help in designing the plants. King, a Dartmouth graduate, studied also at the Zurich Institute of Technology and then became a plant manager for Hoffman-La Roche in Germany. He returned to America following the outbreak of war in August 1914.

During the 1900–20 period Edison became involved not only in chemical problems relating to his storage battery, materials for phonograph records, cement manufacture, and production of organic chemicals during World War I, but he also worked in the field of organic chlorine compounds. Aylsworth played a significant part in this development work, which was done in the Edison Laboratories.

In 1909 Aylsworth patented chlorinated naphthalenes for impregnating wood, paper, and textiles *(13)*. Apparently 1-chloronaphthalene, a liquid, was primarily used. In 1911 Edison patented a composition of approximately four parts of shellac and one part of tetrachloronaphthalene for phonograph records *(14)*. In 1913 Aylsworth covered the solid chloronaphthalenes, with emphasis on the tetrachloride, for various uses as wax substitutes. The Halogen Products Company was formed to market the chloronaphthalenes. The solids were marketed under the trademark of Halowax and were heralded as the first synthetic waxes. This company was originally a part of Thomas A. Edison, Incorporated. Records show that in October 1911 chlorine was being purchased from Hooker Chemical. The monochloride was being supplied to General Electric at Pittsfield in January 1912 in limited quantities. Halogen Products Company later became a part of the Condensite Company with manufacturing operations set up at Wyandotte, Michigan. Here the chlorine was supplied by Dow Chemical, and apparently there was close cooperation between the two companies. A sales pamphlet put out by Condensite in March 1922 stated that the Halowax trademark was used for all of their chlorine products. The Halowax division became a part of the Bakelite Corporation in the merger of 1922 and was included in the sale to Union Carbide in 1939. The Halowax business was subsequently sold to Koppers Company, Incorporated, which company continues to market the various chloronaphthalenes under the Halowax name.

Edison Laboratories studied the synthesis of pentachlorophenol early in 1915. It is possible that this phenol derivative was marketed on a limited scale during the war. It was, and continues to be, a well-known wood preservative. Also laboratory notebook data show that Edison studied the chlorination of rubber in some detail. In an attempt to get away from the expensive process of chlorinating rubber as a solution in carbon tetrachloride, he chlorinated it as thin sheets at an elevated temperature using a blend of chlorine and carbon tetrachloride gas. The latter served to swell the rubber and allowed the chlorine to penetrate and react with it. The chlorinated product was worked up in hot water. Chlorine contents as high as 57.5% were obtained. Magnesium oxide was used as a filler-stabilizer.

Chlorinated rubber has been manufactured in the United States since 1917, but it was not until 1930 that it became commercially successful. Experience showed that it was necessary to chlorinate to a high chlorine content of 66–72% so that the chlorinated rubber would be sufficiently stable for use in paints and the like. Edison's process did not lead to this desired high chlorine content. However, his approach of chlorinating solid rubber was followed up by others. Rubber was ground to a fine powder in proportion of about 1:4 with a salt such as sodium sulfate. The mixture of fines was then mixed with liquid chlorine at 70°–80°F under pressure, allowed to stand for several days, and the sulfate was removed by solution in water. Since this process results in a product of only 60–62% chlorine, it has found only limited use.

Although some phenol is still manufactured by the sulfonation process as employed by Edison, better methods have been developed. A process developed by the Dow Chemical Company in 1924, and still being used by them, is the hydrolysis of monochlorobenzene with aqueous sodium hydroxide. As is well known, chlorine atoms attached to an aromatic nucleus are very stable to hydrolysis. The Dow process requires a temperature of about 750°F at a pressure of 3,800 psi. Adequate pressure must be maintained to avoid vaporization. Aylsworth had basic patents on this synthesis. His patents cover a continuous process using a steel coil reactor at more than 575°F under pressure to avoid vaporization (*16*). One patent covers the process and the other the apparatus. Dow Chemical Company obtained a non-exclusive license under the two patents from the executors of Aylsworth, then deceased, in November 1925 for the sum of $5,000 (17). Thus Aylsworth, and indirectly Edison, had a part in a new manufacturing process which has been considered one of the major accomplishments in chemical engineering in the post-World War I period (18).

During the first two decades of the 1900's Edison had several golden opportunities to set up a continuing chemical business. Even so, he chose to make only certain chemicals needed in his other manufacturing operations. If he had financially backed Aylsworth's developments in the field of phenolic resins as was subsequently done by others, the Condensite Company could have remained in the Edison family of industries. Instead, it fell into the clutches of the Karpen Brothers and subsequently those of the Bakelite Corporation. Today chloronaphthalenes continue to be important chemical products under the original Halowax trademark. Many of the organic chemicals which Edison manufactured during World War I could have been continued with a large growth following the short recession after the end of the war.

Since Edison professed chemistry to be his favorite science, why did he choose not to be a manufacturer of chemicals? The answer was probably age; Edison reached his 70th year on February 11, 1917. Although he had already stated that he was never going to retire, he had said shortly before World War I that he was turning his business operations over to others and concentrating on research. Actually, he continued to "run the show" both businesswise as well as the laboratories, but at this stage in his life he was not looking for new businesses. If his various opportunities in the chemical field had come 20–30 years earlier when he was looking for new industries "to dot the hills" surrounding his new West Orange laboratories, the story might have been quite different.

In May 1915 following the sinking of the *S.S. Lusitania* with the loss of 1,198 lives including 124 Americans, Edison commented that the United States should find some way to retaliate short of war. In an interview with the *New York Times* in late May of that year, he pointed out Germany's intensive use of scientists in preparing for war. With somewhat of a change from his earlier attitude, he stated that if his country needed his services to help make her safe from attack, he would willingly respond. About the same time, Woodrow Wilson and his Cabinet decided that "the inventive, scientific, and industrial power of the country should be mobilized." On July 7, 1915 Josephus Daniels, Secretary of the Navy, wrote a letter to Edison requesting his services as a Government advisor. Excerpts from that letter include:

It is my intention . . . to establish at the earliest moment, a department of invention and development, to which all ideas and suggestions, either from the service or civilian inventors, can be referred for determination as to whether they contain practical suggestions for us to take up and perfect.

We are confronted with a new and terrible engine of warfare in the submarine. . . . I feel sure that with the practical knowledge of the officers in the Navy, with a department composed of the keenest and most inventive minds . . . and with your own wonderful brain to aid us, the United States will be able as in the past to meet this new danger with new devices that will assure peace to our country with their effectiveness . . .

With you it might be well to associate a few men prominent in special lines of inventive research. And I would like also to consult with you as to who these men should be . . . I know the relief which the country would feel in these trying times at the announcement that you are aiding us in this all-important matter . . .

Edison accepted the assignment, and Daniels came to West Orange for further consultation and, no doubt, also to see Edison's research facilities. Edison suggested that rather than the two of them selecting a few men to work with him, Daniels should request each of the leading scientific and engineering societies to select two representatives for this civilian group. This no doubt was a wise decision. The original membership of this proposed department of invention and development was as follows:

Chairman—Thomas A. Edison
American Institute of Electrical Engineers
Frank Julian Sprague and Benjamin G. Lamme
American Institute of Mining Engineers
William L. Saunders and Benjamin B. Thayer
American Society of Mechanical Engineers
William LeRoy Emmet and Spencer Miller
American Society of Civil Engineers
Andrew M. Hunt and Alfred Craven
American Chemical Society
W. R. Whitney and L. H. Baekeland
American Electrochemical Society
Joseph W. Richards and Lawrence Addicks
Inventors' Guild
Peter Cooper Hewitt and Thomas Robins
American Society of Aeronautic Engineers
Henry A. W. Wood and Elmer A. Sperry
American Society of Automobile Engineers
Howard F. Coffin and Andrew L. Riker
American Aeronautical Society
Matthew B. Sellers and Hudson Maxim
American Mathematical Society
Robert S. Woodward and Arthur G. Webster

The chemical profession was well represented by Dr. Whitney, at that time Director of the Research Laboratory at General Electric, and Dr. Baekeland, who was then President of the General Bakelite Company. Since American physicists had no formal organization at that time, Arthur Compton and Robert Millikan, both of whom were later awarded Nobel prizes in physics, were later invited to join the group. Another future member was Lee DeForest, developer of the radio tube.

The first meeting of the group was held in Washington, D.C. on October 7, 1915, at which time the official name of "Naval Consulting Board of the United States" was adopted. At Edison's request he was relieved of the duties of chief executive; William Saunders, a well-known mining engineer, was elected chairman. Edison was given the more-or-less honorary title of president. He could hear little of what went on at the meetings, but his opinions often were solicited by someone's tapping questions to him in Morse code or passing on written notes.

Apparently several members of the Board thought that Edison would be a dominating person and difficult to work with because of his status as a public hero. During the first decade or so of the 20th century, several public opinion polls had voted him the most outstanding living American. As a part of a memorial program sponsored by the American Association for the Advancement of Science at their winter meeting following Edison's death in October 1931, Robert Millikan commented on the impression Edison made on him and other members of the Consulting Board (*19*):

During the war when we were both engaged in Washington I spent an evening or two with him. He was then, at the age of seventy and more, reading some of the newer books that were then appearing in the field of pure science, and asking intelligent questions about them, too. His ears were gone, but there had been no crystallizing of his mind, such as occurs with some of us before we are born; with others, especially with so-called men of action, before we are forty; and with most of us, even with those who have learned to combine the art of knowing with the power of doing, by the time we are seventy. That Edison above all men retained his essential modesty, simplicity, intellectual honesty and willingness to learn, in spite of the disease to which it was his misfortune to be exposed in early life and continuously thereafter, is, I think, the best proof that we have of his real greatness. I refer of course to the disease which attacks and almost always lays low football heroes, movie

Courtesy Edison National Historic Site, West Orange, N. J.

Picture of the Naval Consulting Board taken in Washington, D.C. Miller R. Hutchinson, Edison's chief engineer, is on Edison's right and Secretary Daniels on his left. Undersecretary Franklin D. Roosevelt stands next to Daniels on the front step.

stars, presidents and kings—the disease of publicity and adulation.

Did it go to his head, as it has done in the case of so many others who have been almost great? I could see no indication of it. The fact that his name became probably the best-known one in the United States, that commercial electrical companies from one end of the country to the other adopted it as their trademark, as a guarantee of their quality, that sycophants and exploiters swarmed about him—all this apparently left him practically untouched.

Nor was this merely the impression which he made upon me, one individual scientist. I well remember when Professor Fabry, of Paris, the head of the allied scientific mission sent to the United States by the Allied European Governments during the war, came back from a visit with Edison and said, "Simple, direct, intelligent, unspoiled—a very much greater man than I expected to find in view of the way his name has been exploited and the kind of influences with which he has been surrounded."

The individual members of the Board worked in liaison between the membership of the society which each represented and the Chiefs of the various departments of the Navy. Committees attempted to review suggestions in their respective fields of responsibility. Dr. Whitney was chairman of the committee on chemistry and physics. Thousands of would-be inventors swamped the Navy Department with ideas, drawings, and models. The Naval Consulting Board found, as did various Government technical committees in World War II, that ideas coming from those who have not had specific experience in their fields of suggestion, are largely useless. The principal value of the Naval Consulting Board was that it provided consultants per se to Navy officials, and the Board members arranged for critical research to be carried out in already established research institutions. Secretary Daniels' original thought that Edison's group would act as a clearinghouse for "all ideas and suggestions, either from the service or civilian inventors," was of minor consequence.

Edison was astonished to learn that the Navy did practically no research. Improvements in ship construction occurred in a haphazard fashion largely from ordnance manufacturers and other suppliers. One of the early actions of the Board was to recommend erection of a $5,000,000 research facility to be operated by civilians under the Secretary of the Navy. This recommendation was later carried out but not until after the war.

Late in December 1916 Daniels called on Edison and told him in confidence that according to the British Admiralty, the Allies were fated to lose the war unless the German submarines could be stopped (20). Edison agreed to spend full time in attempting to devise antisubmarine devices and to solve related problems. He dropped all private research he was doing himself and made about 40 members of his staff available for government work. He also recruited volunteers from various colleges and universities and prevailed upon certain industrial concerns to assign some of their technical personnel to the Edison Laboratory to help in the work. The effort became more intensive after February 1, 1917 when Germany declared unlimited submarine warfare on all ocean shipping. After the sinking of several American ships, the United States entered the war on April 6, 1917. During 1917 and 1918 Edison was to spend nearly all of his time on naval research. This was more to his liking than serving as an executive on an advisory board. He cancelled his working Florida vacations in both 1917 and 1918, as well as a jaunt with Ford and Firestone scheduled for the summer of 1917.

In connection with this defense project, Edison and his associates submitted 42 developments and plans to the Navy. Because of the confidential nature of the work, few records remain. Bryan reported on two devices on which Edison spent a great deal of time, including testing at sea from New London in a vessel made available to him by the Navy (21).

Edison on board the *USS Oregon* is greeted by Rear Admiral
William Fulton and Officer Joseph Reeves

Listening Device for Detecting Submarines—This took the
form of an outrigger to be suspended from the bow of a merchant ship. The listening device proper was about twenty feet
long and sixteen inches wide, with a brass body containing
tubes of brass and a phonograph diaphragm at the end that
hung in the water. With a worm worked by an electric motor,
bowsprit and arm could be swung toward the ship, and the
listening device could thus be landed on deck so that necessary
repairs could conveniently be made. A compensating arrangement cancelled out the noise of the ship's engines, and by
aid of an adjustment, confusing noises made by other boats
could likewise be excluded.

Even in the roughest seas, with a ship going full speed (in
that case, fourteen knots), this device resisted injury. While
the ship was proceeding at full speed ahead, other boats could
be heard at a distance of 1,700 yards and a submarine bell could
be heard at a distance of five and one-half miles in the midst of

a heavy storm. It was stated that with this contrivance, a torpedo—"the noisiest craft that sails the sea"—could readily be heard at a distance of over 4,000 yards.

Method for Quick Turning of Ships—This was to be used in connection with the listening device. With this, if the noise of a torpedo had been heard, a merchant ship could quickly change to a course at a right angle to its previous course and thus avoid the torpedo. Four sea-anchors were used. A sea-anchor is a stout canvas bag of conical shape, with a small rope attached to the little end and a heavy rope fastened to the mouth end. This heavy rope is made fast to the ship. Such an anchor is ordinarily used for arresting the speed of a vessel. It is thrown into the sea and, filling with water, acts as a drag. Tension on the small rope turns the device around to empty and retrieve it.

The four sea-anchors used by Edison were each nine feet in diameter at the mouth end and hitched to a four-inch diameter rope. The ropes were firmly attached to the bow and the anchors were placed amidships. If the listening device detected a torpedo, the anchors were to be cast overboard and the helm at the same time thrown hard over. This method was tried successfully with small boats and also with the 5,000-ton U.S.S. "Clio," loaded with 4,200 tons of coal. The "Clio" in two minutes and ten seconds was turned 90° from her original course, with an advance of only 200 feet.

As far as known, the turning technique was never used in actual convoys although it was successfully demonstrated with a tramp steamer. Some aspects of the detecting device were apparently used. The most effective listening device for submarine detection was an electronic one worked out by a team headed by Whitney and DeForest.

Edison felt strongly that more effort should be directed to blinding the enemy's vision *via* their submarine periscopes and less to camouflaging the cargo ships (22). In this connection recommendations were made for smudging the skyline, use of shells containing sulfur trioxide for blinding submarines, and oil for smudging periscopes. Other developments of a chemical nature included a hydrogen detector for use aboard submarines; a steamship decoy which consisted of a floating drum which evolved smoke; and fire extinguishers containing sodium silicate for use on coal-bunker fires.

Karl T. Compton, at that time a member of the physics department at Princeton University and later President of the Massachusetts Institute of Technology, described what it was like to be a member of Edison's laboratory in wartime *(23)*:

Immediately following the declaration of war in 1917 Mr. Edison telephoned President Hibben, of Princeton, requesting him to send to his laboratory four scientists as volunteer war workers. I went with three of my colleagues from the department of physics and remained for the months required to bring to a conclusion the problem which was set for me by Mr. Edison.

All through the war the newspapers published frequent stories of Edison's war activities and of the secrecy in which they were carried out. One story which I remember described experiments carried on in the dead of night on the top of a mountain with armed guards posted all around the base. Whether these stories are true or not I do not know, but I do know that Edison's research laboratory was actively at work and that contact with this work gave me a vivid picture of Edison and his methods.

Immediately upon meeting Mr. Edison and barely taking time to say "how do you do," he took out his pencil and began to describe a problem which had been put up to him by the Naval Consulting Board—the problem of increasing the efficiency of the driving mechanism of a torpedo so that a larger amount of explosive could be stored in it without changing its range or size. He gave me a very brief history of the development of the present torpedo, told me the conditions which an improved torpedo would have to satisfy, and told me to come back to see him when I had a solution.

In about three weeks I reported to him that I had found three fuels which seemed to offer possibilities. He disposed of these solutions in three sentences: "Fuel A can only be obtained in Germany. Fuel B has been tried but discarded because of the danger of explosions. Fuel C, which included wood alcohol, is no good because the sailors drink the d—— stuff."

So I went back for another couple of weeks and returned with a fourth solution. Mr. Edison took the papers, looked over the calculations, muttering the while to himself, and then said, "When I don't understand work like this I get two men to work at it independently. If they agree, maybe it is all right; if they don't agree, I get a third man. Go up into room . . . and see whether you agree with a young fellow from Columbia University whom I put to work on the same problem."

On interviewing this Columbia scientist I found that we agreed entirely as to method but disagreed radically as to conclusions. Whereas I had found very few fuels possibly superior to those which the Navy was using, he had found that almost every fuel was superior. On looking over his work, however, I found that he had based all calculations on a formula for alcohol, $C_{12}H_{22}O_{11}$, which is sugar. In other words, he had been actually finding out what fuels would be better than sugar for driving the Navy's torpedoes. When I asked him where in the world he had got that formula for alcohol he said, "You see, I am a mathematician and not a chemist so I went to the library," and with that he showed me an ancient book on chemistry, in which $C_{12}H_{22}O_{11}$ was actually given as the formula for alcohol.

Following this conference, Mr. Edison arranged for me a visit to one of the naval torpedo stations, where the calculations were checked by the torpedo engineer and the work was left in the hands of the Navy, with what results I do not know.

A second investigation illustrated Mr. Edison's great fertility and imagination. There had been numerous demands for the development of a supersensitive microphone for detecting enemy operations by night or under the ground or beneath the sea. According to Mr. Edison the ordinary carbon granule microphone had too high a resistance, and he wanted to try metal granules, "But," he said, "metal granules are too blamed sluggish. We must make them lighter." He so devised this scheme: First he got a large supply of hog's bristles from a local brush factory; then he plated these hog bristles with a great variety of different metals. Some of them were plated by the electrolytic process, which he used in manufacturing his phonograph records; others were plated by cathode sputtering in a vacuum; and still others by the condensation of evaporated metals. When each of these hog bristles had plated on to it a thin coat of metal, the bristles were cut up into tiny lengths, each about a hundredth of an inch long, by a microtome such as is used in cutting specimens for microscope slides. These tiny little cylinders were then placed in a bath of caustic potash, "the stuff men dissolved their murdered wives in," said Mr. Edison, which dissolved out the hog bristle and left a tiny hollow cylinder of metal, shaped like a napkin ring, and these were the metal granules which were used in place of carbon for the experimental microphones. How well they worked I do not know, since I did not see the conclusions of the test. My guess would be that they did not work as well as carbon, since scientists think that there is a peculiarity in the structure of carbon which makes it particularly effective for microphonic purposes. It was one of Mr. Edison's characteristics, however, that he would not let his own, or any one else's preconceived ideas stand in the way of making a test. This practice certainly led to many futile experiments, but it is equally true that it led to some successful discoveries which caused the scientists to revise their earlier ideas. Edison was not ignorant of what others had done, even though he often appeared to pay little attention to it. A great reader, frequently, before starting, he read everything which had been published on the subject.

Typical of another method of Mr. Edison's work were experiments on flame throwers and on submarine periscopes. The flame thrower was desired to throw a stream of liquid as far as possible. In order to get the right design of nozzle Mr. Edison instructed one of his helpers to build in his shop a whole series of nozzles with every gradation of angle, length and shape within wide limits, and to pick out the one which threw the stream the farthest.

In the case of the submarine periscope the problem was to prevent the deposit from evaporated salt water spray from rendering the periscope mirror non-reflecting. To prevent this several things were tried, one being to bathe the mirror periodically with materials of very low surface tension, which would prevent the accumulation of water in drops. For this purpose a whole series of liquids was tried and the most satisfactory one selected.

These last two war problems illustrate the method of continual search and trial which underlay much of Edison's work. Notable examples are found in his selection of elements for the Edison storage battery and in his preparation of more than 10,000 double chemical salts in the endeavor to find the most satisfactory fluorescent screen for use with x-rays.

It is a mistake, however, to think that all Edison's work was carried on by this search and trial method. Back of everything

World Wide Photos

The Edisons aboard a converted yacht used in testing some of his anti-submarine devices

which he did or tried there was always an idea. The starting point was always the need of accomplishing some purpose, the second stage seemed to be the suggestion of various ways of accomplishing that purpose, and the final stage consisted in trying out these suggested solutions in as thorough and systematic a manner as possible in order to find the best.

Edison's success lay, I believe, equally in his handling of all three of these stages. He was uncommonly alert to opportunities for supplying a need or presenting an improvement. He was uncommonly ingenious in figuring out ways of designing apparatus to do what he wanted it to do, and he was one of the most patient and persevering men who ever lived in carrying through his ideas to the last stage of comprehensive test.

Edison's free-and-easy manner of doing things often aggravated Navy authorities whose operations were based strictly on protocol. Edison might pick up the telephone any hour of the day or night and request Secretary Daniels with some of his experts to meet him the following day to look over some "big development" which had been made. When he was testing on a naval vessel, he would bring Mrs. Edison along even though women on such ships were strictly taboo. Government security officials strongly objected to his employing German chemists in his laboratory, particularly recent immigrants. Edison "solved the problem" by ordering the laboratory put on a three-shift seven-day week basis with some Germans on each shift. "They won't blow up the place," said he, "as long as some of their former countrymen are in the laboratory!"

Following World War I, Edison prepared a memorandum summarizing his ideas on the nature of man and what he felt would be a proper program for the country to follow during the reconstruction period. The introductory part of that memorandum follows (24):

The present war has proven that man is still a predatory animal, selfish and grasping; wearing the mask of civilization to hide his natural characteristics. This statement is proven by the death of two million men on the firing line.

271

In cities, the police curb him, and the general government performs the same function for the larger units.

As to nations, there is no curbing them. There is no international authority to coerce nations; therefore, we should take precautions.

Nearly every nation is predatory more or less, except the United States.

The United States is a rich prize for a predatory nation or combination of nations.

The United States should prepare to defend itself against any possible invasion.

Chapter 9 **Rubber from Goldenrod**

AFTER 90% OF THE world's rubber-growing area was lost to Japan in World War II, the Baruch Committee warned its fellow Americans as follows: "Of all critical and strategic materials, rubber is the one which presents the greatest threat to the safety of our nation and the success of the Allied cause. . . . if we fail to secure quickly a large new rubber supply, our war effort and our domestic economy both will collapse. Thus the rubber situation gives rise to our most critical problem." In summarizing its recommendations, the Committee added, "It appreciates that it is asking the public to make sacrifices because of mistakes that have been made and for which the public is not to blame."

Although the U.S. Government made a minor effort in the 1923–27 period to determine if rubber-producing trees could be grown any place in the Americas, it was largely private enterprise which made some valiant attempts to establish a source of supply closer than the Far East. Henry Ford accepted a recommendation of the government survey that an area in the Amazon valley was suitable for rubber growing. He established two plantations along the Tapajos River, but fungus diseases and pests were victorious over a major and expensive effort lasting for 18 years.

Harvey Firestone relied on a survey by his own experts as well as one by the Government when in 1924 he chose Liberia

Courtesy Edison National Historic Site,
West Orange, N. J.

Edison receives an honorary Doctor of Science degree at Princeton University. Dean West leads the way.

for locating his rubber plantations. This African country is closer to U.S. markets than the Far East. Firestone's foresight provided a small but significant source of natural rubber for the United States during World War II.

In the early 1920's the Intercontinental Rubber Company attempted to establish guayule as a source of domestic rubber with plantations both in Mexico and southwestern United States. Although rubber from this source was of satisfactory quality, it was not competitive price-wise with *Hevea* rubber in peacetime. A few chemists talked of synthetic rubber, but the rubber fabricators unanimously agreed that a general purpose synthetic rubber could never compete with natural rubber in a free market.

However, there was one individual who as early as 1920 grasped the seriousness of relying on a source of rubber grown half-way around the world from our processing plants. That person was Thomas A. Edison. Unlike others who attempted to produce alternate sources of rubber supply, Edison saw little chance of competing economically with the favorable growing conditions and, above all, the cheap labor of the Far East. Rather he visualized that the best solution was to develop a source of rubber from plants which, in an emergency, could be grown and harvested within a matter of months. Under such a situation, production cost would not be a significant factor. Edison was well on the way to attaining his objective when in 1931, in his 85th year, death came. Synthetic rubber ultimately provided the domestic source of supply which he sought and, in addition, proved to be competitive with natural in peacetime.

Columbus, in describing one of his voyages to the New World, referred to "bouncing balls used by Haitian natives." Archeological investigations of Mayan culture have confirmed that rubber balls were used at least as early as the 11th century. In 1735 members of the French Academy of Sciences obtained samples of rubber from natives of the Amazon valley and reported on the origin of this elastic material. The natives called it, cahutchu, derived from caa, meaning wood, and o-chu, meaning tears. In French the name is still caoutchouc, and in German, kautschuk. The famous English scientist Joseph Priestley observed that when this natural gum material was rubbed over pencil markings and such, they were erased; hence he named it rubber. With the advent of many rubberlike substances prepared synthetically, this broad class of materials is now known scientifically as elastomers. However, the term rubber continues to be used broadly for both natural and synthetic rubbery materials. The American Society for Testing Materials defines rubber as "a natural or synthetic material that can be, or is already, vulcanized to a state in which it has high extensibility and forcible, quick retraction." In other words, a vulcanized rubber will stretch and snap back when released. We will discuss vulcanization later.

Natural rubber is obtained as a white milky liquid, called latex, from certain types of plants when the surface is cut. It originated in the tropical jungles of South America and Africa. Latex contains 30–50% rubber dispersed as fine particles in the aqueous medium. Based on total solids, the hydrocarbon rubber constitutes 92–94%, with the other 6–8% consisting of resins, proteins, and other complex bodies. The non-hydrocarbon materials keep the rubber dispersed and protect it against degradation by air, sunlight, and the like. The early practice for recovering rubber from latex was to dip a stick into the latex and then heat it over a fire. As a result of the temperature rise the water evaporated and the solid rubber coagulated.

Of the various species of plants which produce the latex, the *Hevea* tree proved best, and seeds of the *Hevea brasiliensis* were brought by the British to London in 1873, and young trees were grown in Kew Gardens there. These trees were taken to British-controlled lands in southeast Asia in 1876 and successfully grown in Malaya and Ceylon. These were the start of the vast natural rubber plantations which are now concentrated in that part of the world. Whereas in 1899 only 4 long tons of plantation rubber were produced, in 1970 production was 2,912,500 metric tons. [A long ton equals 2240 lbs; a metric ton equals 2204.62 lbs (1000 kg).]

In 1860 an English chemist, Greville Williams, heated rubber in the absence of air and found that it broke down to give tarry materials and a low boiling hydrocarbon. The latter proved to have an empirical composition of C_5H_8 and was given the name isoprene. In 1875 Bouchardat advanced the idea that isoprene is the primary unit of rubber, and he succeeded in producing a rubbery polymer by heating isoprene with fuming hydrochloric acid. Subsequent work by many scientists over the years has proved that natural rubber is a polyisoprene having repeating units of

$$-CH_2-\underset{\underset{\displaystyle CH_3}{|}}{C}=CH-CH_2-$$

The formula for rubber can be expressed as $(C_5H_8)_x$ where x may vary from a few hundred to 100,000 or so. Thus, rubber has huge molecules with molecular weights up to a few million.

The natives of the Amazon used rubber to make balls, bottles, and shoes. A bottle-shaped clay mold was covered with latex and allowed to dry in the sun. When the gum solidified, the clay mold was broken, and a container for liquids much lighter than those made from earthenware resulted. The first commercial factory to manufacture rubber goods was established by Charles Macintosh in Glasgow in 1823. Waterproof fabrics bearing the inventor's name were made by applying a solution of rubber in a solvent, such as benzene, to fabric. Rubber shoes were made in Roxbury, Massachusetts as early as 1833. However, as with any article containing rubber, they hardened in cold weather and became tacky in hot weather.

Two discoveries led to improvements whereby rubber became an important material in commerce. Thomas Hancock, a London coach builder, found that if crude rubber is mechanically torn apart or kneaded between heavy metal rolls, it becomes plastic and assumes the consistency of a stiff dough. In this form, powders, including sulfur, waxes, resins, and oils can be incorporated. Mastication of the rubber under shear

277

actually breaks down the big molecules from several million molecular weight to something on the order of 50,000–100,000. In 1832 F. Ludersdorff in Germany noticed that rubber which contained a few percent sulfur lost its tackiness if heated. However, he apparently did not realize the importance of his discovery. In 1838 Charles Goodyear in the United States found the same thing. After learning that Nathaniel Hayward had reduced the surface tackiness of rubber by sprinkling powdered sulfur on it and exposing it to the sun, Goodyear undertook a study of heating rubber mixed with various proportions of sulfur. Using superheated steam he found that such mixtures could be converted to a dry, tough, highly elastic material. A temperature of at least 300°F is required for the reaction of sulfur and rubber to proceed at a rapid rate. This chemical process became known as vulcanization. Today—over 130 years since Goodyear's original discovery—sulfur is still the most widely used chemical for curing both natural and synthetic elastomers.

Goodyear learned that vulcanization could be accelerated by adding other chemicals such as zinc oxide (ZnO), litharge (PbO), or lime (CaO) to the rubber mix along with the sulfur. Rubber articles could then be made which were strong, non-tacky, and which retained their strength at much higher temperatures than previously. By 1849 there were 36 rubber manufacturing establishments in the United States making rubber footwear, belting, hose, and numerous other articles. Solid rubber tires were first used for carriages in 1857, and the "bicycle age" beginning in about 1890, led to pneumatic tires. The motorcar industry, ushered in at the turn of the century, made the production of rubber and the manufacture of rubber articles an important part of world commerce.

In Charles Goodyear's day rubber sold for 5¢ per lb. In 1849 the United States consumed 1,000 tons. By 1900 United States factories used 20,000 tons, and the price had increased to about

60¢ per lb. With the Model-T Ford coming off production lines in the first mass production of any automobile, demand for rubber for tires skyrocketed, as did the price—$2.15 per lb in 1909. The first plantation rubber reached London in 1905. Although the shipment was only 174 tons compared with Brazil's 40,000-ton annual shipments of wild rubber, the era of plantation rubber had arrived. By 1910 plantation rubber production reached 11,000 tons, and by 1914 it had outstripped wild rubber, and the price of rubber in New York was 65¢. For the next decade or so, the production of plantation rubber kept pace with or exceeded the demand.

Following World War I there was a world surplus of rubber. This plus the business recession in the United States caused rubber to fall to 16¢ per lb in the autumn of 1920. The following year the price fluctuated between 12 and 21¢, and in 1922, 14 and 29¢. These prices were lower than production costs at that time. The British growers tried a voluntary agreement to cut back on production and thus raise the market price, but this failed. In November 1922 the British Government through its Secretary of State for the Colonies promulgated the Stevenson Rubber Restriction Act, which was designed to control the sale of rubber to avoid surpluses which depressed prices. The Dutch did not join the British in their control scheme, and production increased in Java and Sumatra while it was being curtailed in Malaya and Ceylon. The Stevenson Act did, however, serve to bring the price of crude rubber up to the 30–35¢ range.

The United States at this time consumed over 70% of total world production, largely for use in tires and tubes. In 1925 the demand for rubber drastically increased in the United States, largely because of a shift to balloon tires which required more rubber per tire than the old high pressure type. By July of that year the United States was down to one month's supply of rubber, and the price had shot up to $1.23. These chaotic conditions caused many to look into the possibilities of producing rubber in North America, or at least somewhere under American control.

Harvey Firestone was the leader in fighting production restrictions abroad and moved to grow rubber to meet his own requirements. His experts surveyed conditions in the Philippines, Liberia, Mexico, and Panama among others. Liberia was chosen as the most likely site, and a *Hevea* plantation of 1,100 acres at Mount Barclay which had been abandoned by the British in 1918 was taken over. This overgrown area was cleared of brush and was operated at a profit in 1925. Negotiations with the Liberian Government led to the leasing of up to 1,000,000 acres in 1926. A total of 20,000 acres were cleared, and seedlings were planted shortly thereafter. This project made slow headway because of low prices of Far East rubber following the repeal of the Stevenson Act in 1928 and the severe depression of prices during the 1930–1935 economic recession. However, Mr. Firestone's foresight was a significant factor during World War II when Liberia became the primary source of natural rubber to the United States since the British took most of Ceylon's production, the only other major source not captured by the Japanese. Natural rubber production in Liberia has continued and by 1964 had reached a total of 244,724 acres of rubber trees on estates and native plantings. In 1970, Liberia furnished 7.5% of the total world's natural rubber production.

By 1925 the automobile industry had overtaken that of steel in size and was second only to agriculture. Henry Ford realized that rubber tires were essential for his Model-T's and that a stable moderate rubber price was essential for one to run an automobile business. Tires represented more than 10% of the original cost of the Model-T, and because of the low mileage (5,000 or so) also represented the major maintenance cost. In 1923 Congress appropriated $500,000 to carry out a study on likely locations for growing *Hevea* trees. Surveys were made in Latin America, the Philippines, and our southern states. Ford accepted a recommendation of the Department of Commerce that the Tapajos Plateau in Brazil was a likely spot. His first plantation failed because of the South American Leaf Blight, a virulent fungus disease. He moved his operations to a plateau area lower on the Tapajos, but again the leaf blight was a serious problem. It was overcome by top-building the high yielding trees with tops of disease-resistant trees growing wild in Brazil. However, this increased production costs. The Ford plantation holdings there were sold to the Brazilian Government for a nominal sum shortly after World War II.

As a result of the surveys carried out by the U.S. Department of Agriculture relative to likely areas for growing rubber, plantings were initiated in southern United States, Haiti, and Panama. However, these projects were abandoned during the economic crisis of the early 1930's, when the price of crude rubber fell below 3¢ per lb.

Edison, Firestone, and Ford were great friends. Edison's and Ford's winter homes were adjacent to each other on McGregor Boulevard on the bank of the Caloosahatchee River in Fort Myers, Florida. Edison's second wife, Mina Miller, came from a prominent family in Akron, Ohio, the city that was already known as the "Rubber City" when Harvey Firestone settled there in 1900. All three men were highly individualistic and leaders in their fields. Each appeared to follow the common motto, "Go it alone. Do not fail to try because someone has already tried and failed."

Starting in 1918 the three made numerous summer camping trips together. Since these extended through 1924, there is no doubt that the subject of the availability of rubber to the American manufacturer was a frequent topic of conversation. On their camping trip in 1919, Firestone relates that he had a lengthy discussion with Edison about rubber and its properties

(*1*). Firestone, surprised at Edison's knowledge of the subject, relates "he told me more than I knew and more than I think our chemists knew (about rubber), although to the best of my knowledge, Mr. Edison had never given any attention to rubber, except in connection with his talking machines."

As industrialists, Firestone and Ford naturally turned to the then current method of producing rubber from the *Hevea* tree but under conditions more favorable to them and to the country. Edison, the innovator, sought other approaches. He was apparently satisfied to let his two friends worry about price stabilization and getting enough American production to avoid the British and Dutch monopolies. His primary interest was in a source of rubber during an emergency. He had seen many strategic chemicals and their derivatives shut off during World War I, but by proper know-how these were made available by American manufacture. He resolved that the United States should be able to do the same in case of rubber, even though such a source might not be practical during peacetime.

In 1926 Edison was no doubt well aware that rubberlike materials had been made synthetically. By 1901 it was discovered that diolefins other than isoprene could be converted to rubbery materials. In fact, during World War I the Germans had made three grades of rubber from dimethylbutadiene. This conjugated diolefin, the next higher homolog above isoprene in this chemical family, was synthesized from acetone, a low boiling liquid available at that time from the dry distillation of wood or from the pyrolysis of calcium acetate. The acetone was first reduced with a metal, such as amalgamated magnesium, to pinacol, tetramethylbutylene glycol. Subsequent dehydration gave 2,3-dimethylbutadiene:

$$CH_2{=}C{-}C{=}CH_2.$$
$$\underset{H_3C}{|}\ \underset{CH_3}{|}$$

As president of the Naval Consulting Board, Edison was well

aware of the technological problems faced by the Germans. He was also no doubt aware of the unsuccessful work done by Thomas Midgley, Jr. of the General Motors Corporation on synthetic rubber during the early and mid 1920's. Midgley and his associates had primarily studied the polymerization of isoprene using alkali metals, such as sodium, as catalysts (2). They had found that the purity of the isoprene used was very important. Solid rubbery products obtained did not have the "nerve" of masticated rubber. Pure gum vulcanizates prepared with sulfur, an organic accelerator, and zinc oxide as curatives had tensile strengths in the range of only 600–1800 lbs per sq inch, as compared with 4,000 or so for like vulcanizates of natural rubber. Since isoprene was available only from suppliers of reagent chemicals at that time, huge chemical facilities would be necessary to prepare the raw materials for manufacturing synthetic rubber.

On the other hand, Edison had the basic philosophy that Nature provides answers to problems if one will only seek. Since 1923 when he first became interested in an alternate source of rubber for the United States, he had undertaken the study of botany with emphasis on those families of plants which were known to contain some rubber. Much had been published on the subject. As a result of the joint survey carried out by the Departments of Commerce and Agriculture using the Congressional appropriation of 1922, most domestic and many foreign plants known to contain milky juice or sap had been analyzed for rubber. Among others, these included milkweeds, Indian hemp, wild lettuce, sumac, and hedge balls known to every midwest boy to contain a sticky milky liquid. One investigator stated in November 1926 that "The number of plants known to produce rubber is upwards to one hundred, and the list is constantly increasing" (3).

At the semiannual meeting of the American Chemical Society (ACS) held in Philadelphia in September 1926, the

Division of Rubber Chemistry sponsored a symposium on "Raw Rubber." The twenty-odd papers presented were later published in the November issue of the applied journal of that society, *Industrial and Engineering Chemistry* (4). These papers covered in an excellent manner the status of rubber production, alternate sources—both natural and synthetic—chemical and physical properties of rubber, and vulcanization. Anyone wishing to know the status of the art of rubber as a raw material at the time Edison was undertaking his studies in this field can profitably read the papers presented at this symposium. Edison no doubt read them since he was an ACS member and subscribed to *Industrial and Engineering Chemistry*.

At that symposium, three papers dealt with rubber from the guayule shrub. This woody plant grows wild in extensive reaches of Mexico and southwestern Texas. It reaches a maximum height of about two feet, and for maximum rubber production a life span of 4–5 years is most favorable; the average plant weighs less than one pound. Although the rubber is present as a latex, the plant does not have a duct system as does *Hevea*. It exists primarily under the bark of the trunk, major branches, and roots, and this impervious sheath appears to be Nature's way of preventing loss of moisture from the plant. The wild shrub contains about 8% rubber on a dry weight basis while cultivated species run as high as 22%.

By 1902 commercial operations to recover guayule rubber began in Mexico, and the first commercial shipment was made to the Manhattan Rubber Company in 1904. The guayule industry produced 8,500,000 pounds of dry rubber in 1925. The principal producer was the Intercontinental Rubber Company with plantations both in Mexico and the southwestern states. Starting with seed, about 7,000 plants per acre were grown. The cultivated shrubs were completely uprooted, and new plantings were started every four or five years. Yield from such a harvest was on the order of 1,600 pounds of rubber per acre. On an annual basis this compared favorably with *Hevea* which had an average annual yield worldwide of about 400 lbs per acre. The recovery operation involved crushing the whole plant in the presence of water and then ball-milling. The rubber particles were then floated off from the slurry obtained with the water-logged cellulosic material settling. The rubber was washed and sheeted on a corrugated roller mill in the same manner as reclaimed rubber. The price of guayule rubber has usually run about 80% of that of *Hevea* rubber.

Although Edison realized that many plants had been found which contained some rubber, he felt that a systematic examination of literally thousands of plants might well uncover some new and more promising type. His visit to Luther Burbank's experimental farm at Sebastopol, California, in 1915 demonstrated to him that plants could be modified and desired characteristics improved by crossing and selection. Thus, finding the best plant to serve as a domestic source of rubber production would be only the first step. Next he would improve it as Burbank had done with the potato, various berries, fruits, and flowers. Edison already had considerable know-how in growing exotic plants at his winter home in Fort Myers, Florida, and he had good relations with various botanical gardens. Although guayule showed considerable promise at that time as a possible domestic source of rubber, Edison felt that the Intercontinental Rubber Company was doing an adequate job in this field. Further, his objective was not to establish a domestic source of rubber which would be economically competitive with rubber plantations in tropical countries having abundant cheap labor. His objective was a source of domestic rubber in case foreign supplies were shut off by war or severe production curtailment from *Hevea* as a result of plant diseases or fungi attack. This meant a plant which could be grown rapidly, preferably more than one crop a year, and which could be harvested and processed largely by machinery.

As early as 1923 Edison began to experiment with milk-weed and Mexican guayule plants which had been uprooted and shipped to his laboratory at Fort Myers. By 1925 his interest had shifted to *Cryptostegia grandiflora,* a species of Madagascar vine which could be grown in the South. This plant contains rubber in all its parts except its leaves.

By the time Edison had returned to his New Jersey home and laboratories in the spring of 1927, his primary research project had become the collection and analysis of plants for their rubber content. He suggested a revival of the Firestone-Ford-Edison camping trips primarily to collect plants for his tests. Ford was too busy because of his work on a new car to replace the Model-T, but he was delighted to know that Edison was going to concentrate on the rubber problem. Based on previous discussions, a small company would be formed to handle this research and development project. Ford suggested that he and Firestone contribute the necessary money and Edison his services. Thus the Edison Botanic Research Corporation of Fort Myers, Florida came into being in 1927. It is not entirely clear how much money Ford and Firestone put into the project but it appears to have been $25,000 each. Edison furnished the manpower and facilities, and since the work went on for over four years, he also provided the major financial support. His close relationship with Firestone was very helpful in developing techniques for evaluating experimental samples of rubber. In cooperation with George Carnahan, the energetic president of Intercontinental Rubber, Edison purified several hundred pounds of guayule rubber by extraction and turned this over to Firestone for making tires for his Ford truck. This vehicle was used at Edison's experimental gardens in Florida and for searching the Florida woods for new plants to test. Edison reported that the tires wore well but cracked badly, including the seldom-used spare.

We now know that Nature does not provide the highly effective antioxidant materials in guayule rubber as she does in *Hevea* rubber. Furthermore, during the purification of the guayule rubber, the impurities which did protect the rubber to some degree were largely removed. The cracking of vulcanized rubber is caused by the chemical reaction between the rubber and ozone in the air and is accelerated when exposed to sunlight. Present-day technology has provided additives which are incorporated into the rubber to prevent this chemical attack.

In the summer of 1927 Edison engaged several botanists to collect plants and send them to the West Orange Laboratories. The following letter was sent to a prospective plant collector and typifies Edison's thinking in initiating the project (5):

I am engaged in an investigation looking toward the production of rubber in the United States from plants, bushes, shrubs, etc. which can be planted by acreage and harvested with reapers. To this end I desire to collect and examine as completely as possible the various species of the plants, bushes, shrubs, etc. which contain a rubber-bearing latex, and to accomplish this result I intend to engage several field men who have specialized in botany. The compensation would be one hundred dollars a month and expenses, and the field to be covered would be in Southern States up to as far north as New Jersey. The term for this work would be in the summer months.

On July 1, 1927, Edison sent a second letter in which he emphasized the importance of latex plants and particularly the Euphorbias:

I am enclosing the instructions for collecting latex-bearing plants, in accordance with our previous correspondence; which please read carefully. . . .

I am also sending to you an incomplete list of the Euphorbias. It is the Euphorbias that promise to be the best kind of plant for my purpose.

I have stated in what places the plants are found wherever I have been able to find them in the sources of my information. The United States has not been thoroughly explored, and I think that many plants mentioned as growing in foreign countries will be found also in the United States. Lists of the Asclepiadaceae, Apocynaceae, and other families will be sent to you in the near future as soon as I get your first address on the road. None of these are as promising as the Euphorbiaceae.

For each one of the plants received by me from you men in the field, a herbarium card will be made, and identification will be sought by comparison at the herbaria of botanical gardens in Washington, New York and other places.

We are fortunate in having one criterion that is certain, and that is, the latex flow on cutting.

If a new species is discovered, the finder will get the credit.

Thus in the initial studies, plants containing milky juice were emphasized. Accordingly, Edison pursued the *Euphorbiaceae,* commonly known as the spurge family, since *Hevea* belongs to that family of plants. As a home owner in east central New Jersey, the author can testify that spotted spurge is a most persistent weed in this area. It is little affected by the usual weed killers, and long before I knew Thomas Edison had worked with spurge as a possible source of rubber, my sticky hands resulting from pulling this weed from my suffering lawn attested that here was a potential source of rubber.

In July of 1927 William Meadowcroft, Edison's private secretary, wrote to the various field collectors as follows (5):

Mr. Edison desires me to request that you will not collect specimens of the ordinary variety of milkweed, nor of sumac, thistles or wild lettuce. We have an abundance of these around here.

So long as the plants give latex and are not of the kind above specified, there need be no special delay by reason of your trying to identify the plants collected. We shall send one out of each bundle of your specimens to the New York Botanical Gardens for identification.

On July 15, Mr. Meadowcroft wrote as follows:

In pursuance of Mr. Edison's direction, I am sending to you herewith a copy of the Bulletin (323) issued by the U.S. Department of Agriculture, entitled "How to collect, label, and pack living plant material for long-distance shipment." Mr. Edison says you will find some very useful information in this bulletin.

In packing and shipping your collection of plants, let me once more emphasize that you should be careful not to cover up the leaves too closely with packing. Any packing around the leaves should not go all the way up and should be light and very loose. The leaves of a few specimens so far received (not necessarily from you in particular) have rotted and fallen off.

Tests soon showed that the presence of latex—*i.e.*, a milky juice—was not necessarily an indication of rubber. Thus, later in 1927 the program was broadened to include all plants, with emphasis on those families where encouraging results had been obtained with at least one species.

Dr. Loren G. Polhamus in his excellent compilation of Edison's work on rubber from plants describes further the sources and number of plants collected (5):

In addition to the collections made by his own agents, Mr. Edison also received plant specimens from volunteer collectors who heard of his efforts and wished to contribute to their success. A major contribution to the success of the surveys was made by the Union Pacific Railroad, officials of which, at the request of Edison, directed each section foreman to make collections of locally available plants along the right of way and forward them to the Edison laboratories for tests. This resulted in the receipt of thousands of samples of plants, from areas not covered by Edison's own collectors and thus giving him representative plants from a much wider area.

In all, some 17,000 samples of plants were received at the Edison laboratories in West Orange, N.J. Despite the careful instructions given to all collectors, many of the samples were received in a decaying condition. Others lacked flowers, fruit, roots, or other plant parts needed for identification. A total of 13,344 specimens were identifiable at least as to genus. There were, of course, many duplicates, but a total of 2,222 species were represented. These were classified in 977 genera and 186 natural plant families. This represents by far the largest group of plant samples collected and tested for rubber content by any investigator. It still outnumbers the total of all other investigators . . . prior to Edison's collections and, after his work . . . in the United States and . . . Canada.

When a plant sample was received at the laboratory, an herbarium specimen was taken and preserved for botanical identification. Such a specimen is a dried sample where an attempt is made to preserve parts of all portions of the plant. Once a considerable number of specimens had been collected, they were submitted to Rutgers University for identification, using Britton's State Herbarium (an outstanding collection) which was stored there, for comparative purposes. After Edison's death in 1931, the herbarium specimens were boxed and shipped to the U.S. Department of Agriculture in Washington where identifications of all specimens were rechecked by Sydney F. Blake. During World War II when Edison's work was reviewed and extended by the U.S. Forest Service, a further check of the identifications and nomenclature was made by Doris Hayes of that department.

In testing a plant for rubber content, the entire plant might be used or some specific parts of it, such as leaves only, the stem, etc. This was so designated in the laboratory records. The method used for screening the various test plants was an extraction technique with only qualitative evaluation of the product. Polhamus outlines the initial testing procedure:

The specimen to be tested for rubber content was first dried in an oven and then ground. A test sample was next weighed out and the remaining powder retained for future use if needed. The test sample was placed in a Soxhlet extractor for rubber determination. Each sample was first treated with acetone to remove the nonrubber constituents of the plant that might be soluble in benzol. This acetone extract was designated as "resin" and discarded without further examination and without any record of amount. After the extraction with acetone, the sample was first freed of acetone and then extracted with benzol. The benzol extract was then placed in an open dish and the benzol evaporated off. The extract was weighed and designated "rubber."

Edison took a very active part in the overall rubber project. He examined all collections and had Dr. J. K. Small, a noted botanist who was hired during the summer months, identify and give each specimen its botanical name. He required that all dried benzol extracts be placed on his desk for his personal examination. This might total as many as 50 a day. He made notes as to color, degree of tackiness, and whether the elasticity was good, fair, or poor. If good, he recorded whether the sample was strong or weak and if long or short on elongation. If a sample indicated a relatively high proportion of good rubber, the collector was requested to get additional samples in the same area.

To the chemist trained in the science of quantitative analysis, this testing procedure may appear to have been extremely empirical. However, even today the "feel-and-bite" test of gum rubbers can be a very meaningful evaluation in the hands of an expert. The physical properties of a polyisoprene rubber are more important than its chemical analysis since low and high molecular weight materials have similar chemical analyses whereas their properties in a vulcanized article may be vastly different. Edison's long experience with materials for phonograph records made him highly skilled in evaluating polymeric materials by hand examination.

Acetone has long been used as an analytical tool for separating nonrubbery materials from natural rubber by Soxhlet extraction. All grades of natural rubber are soluble in the aromatic hydrocarbons, and benzene is the best solvent in this chemical family. Cellulose is insoluble in both acetone and benzene. Thus, if a plant tested as cited above did contain rubber, that rubber would end up in the benzene extract. On the other hand, just because a benzene extract was obtained, that did not necessarily mean rubber was present. Some extracts were marked NR, meaning no rubber. In the case of other extracts, Edison would request that they be purified. This was done by re-solution in benzene, followed by coagulation with acetone. The impurities would tend to remain in the benzene solution and the rubbery "gunk" would be washed with acetone and then freed of solvent by heating.

Edison recognized early that the rubber in a plant might be degraded by oxidation during the recovery process. The fact that the benzene extracts, even those high in rubber, were much more tacky and "gunky" in nature than either *Hevea* or guayule rubber indicated possible degradation. Also, a sample that was purified by redissolving and precipitation might be less elastic than the original. The analytical procedure was modified by drying a plant at 120°F rather than at 212°, by extracting in an atmosphere of carbon dioxide, and by removing the acetone from the Soxhlet and the benzene from the final residue under mild temperature conditions under vacuum. Results were largely inconclusive. However, the factor of possible oxidative degradation during recovery as well as the desire to use a more quantitative analytical method led to the use of the following procedure as pointed out by Polhamus:

In the latter stages of Edison's plant surveys and in most of the tests of cultivated plants from his experiments at Fort Myers, the method of determining rubber content by double extraction with acetone and benzol was abandoned and a bromination method was adopted. In this method, benzol was used to extract all solubles (in benzol) from the ground sample. A solution of bromine in carbon tetrachloride was added to the extract to transform the rubber portion to rubber bromide. The bromide was then precipitated with alcohol and washed on a vacuum filter with additional portions of alcohol to separate it from the nonrubber substances extracted by the benzol. It was determined empirically that the rubber bromide was 28.5 percent (0.285) rubber and this factor was used in converting dried rubber bromide to rubber for use in calculating the rubber content of the sample.

This method was used largely for plants known to contain

rubber, and thus it was unnecessary to isolate a benzene extract for physical examination.

C. A. Prince in late 1929 worked on a direct bromination process for determining the rubber. The dried plant was ground to pass a 30-mesh screen and then was washed successively with 0.25% sulfuric acid, water, 0.25% sodium hydroxide, and water. The powder was then dried and slurried in benzene. After an hour or so of contact, an excess of bromine in carbon tetrachloride was added; the bromine which did not combine with the rubber was determined by adding excess potassium iodide solution, and the liberated iodine was titrated with standard sodium thiosulfate solution using starch solution as indicator. The bromine test was also used directly on green leaves without drying or grinding.

Edison and his associates studied many modifications of the extraction technique. For those plants containing considerable tannin, it was found advantageous to slurry the ground material in aqueous sodium hydroxide, water wash, and dry prior to the acetone-extraction step. Edison disliked the use of benzene because of its known toxicity and high flammability. Chlorinated hydrocarbons are also good solvents for rubber, but many of these are also toxic. For a time blends of benzene and chloroform were used. Work with dichloroethylene looked promising, and Edison recorded in his notebook of July 9, 1927, "This gives us 2 strings in our bow." However, benzene continued to be the best overall solvent and had the additional advantage of costing only 22¢ per gal *vs.* $2.08 for the dichloride. Edison noted in his notebook of July 17, 1928 that benzene manufacturers apparently handle the solvent without accident. Laboratory workers were instructed to evaporate the benzene from extracts outside the laboratory when at all possible. A favorite spot to do this at the Fort Myers laboratory was in an adjacent garage. C. A. Prince, Jr., a personal assistant to Edison the last year or so prior to the latter's death, feels that adequate precautions were taken to avoid poisoning of laboratory personnel by benzene and other toxic solvents handled in the rubber work (6). He cites the record of his father, C. A. Prince, who probably worked with benzene more than any other person and showed no signs of systemic poisoning up to the time of his death. Considerable credit should be given Edison for recognizing the potential hazards of benzene handling since it has only been the last 15 years or so that toxicologists have had data to show that benzene is a much more toxic material than other aromatic hydrocarbons, such as toluene and the xylenes.

The laboratory work on rubber was carried out at West Orange during the summer and fall months with the field people sending in plant samples of various annuals and perennials. Many who read about Edison's project sent him plant samples, particularly those containing a sticky sap. Literally hundreds of hedge balls were sent in by unsolicited helpers. Samples of rubbers were received from self-styled inventors with requests that Edison test these promising materials. In Edison's notebook of October 5, 1928, there is a notation relative to a sample of synthetic rubber sent in by one H. A. Freeman of Connecticut. Ignition had shown that it contained 54.5% ash, most of which was zinc oxide. At the bottom of the page: "TAE Fraud!"

During winter months the rubber work was transferred to Fort Myers where a special laboratory had been built for the studies. Nine acres of Edison's gardens were turned over to grow promising plants for rubber testing. Work groups, including Edison himself, would make foraging trips into the Florida woods and fields near Fort Myers seeking new specimens for testing. By 1929 Edison had announced that goldenrod was the best of the thousands of plants tested, with a rubber content of about 4% in the leaves. Intensive studies a la Burbank began with the *Solidago rugosa* and *Solidago leavenworthii* species of the goldenrod. They were planted in experimental

Courtesy Edison National Historic Site,
West Orange, N. J.

Experimental plots of goldenrod at the Edison Botanical Gardens, Fort Myers, Florida

plots at Fort Myers in January and harvested from May to November. In addition to cross-fertilization, various fertilizers, best time to harvest, effect of close planting, etc. were also studied. Several acres up the river at Fort Thompson were planted with goldenrod for larger scale recovery studies. In 1929 Edison reported that one could produce 100 lbs of rubber per acre from goldenrod, and he hoped to increase this to 150 lbs. A *leavenworthii* species growing as high as 14 feet was found in the Florida Everglades, and this was crossed with a smaller variety to produce a plant about 12½ feet high with a maximum of 12.5% rubber in the upper leaves, on a dry weight basis.

The leaves are the only part of the goldenrod which contain appreciable quantities of rubber which is present in the solid

295

state as globules in the leaf cells. Rubber accumulates first in the lower leaves and progresses upwards as the upper leaves reach full maturity. The leaves are usually shed soon after full maturity. Thus the shedding process works progressively up from the base of the plant. A harvest of the lower leaves at the time of their maximum rubber content, followed by harvesting the remainder later, would be best.

Edison attempted to develop a mechanical process for recovering the rubber from goldenrod leaves similar to that described above for guayule recovery. The process involved grinding in such a manner that the coarse fibers could be rejected prior to water-logging the pulp and floating off the rubber. The process was never adequately developed.

Rubber obtained from goldenrod is, of course, polyisoprene as is all rubber obtained from plants. It is of low molecular weight and is semiliquid. In order to handle the rubber on a roller mill, it was necessary to mix it with carbon black, and this was done in a ball mill. Since carbon black is used in proportion of 50 parts or more per 100 parts of rubber in practically all compounds, this presented no serious problem. However, difficulties were encountered in vulcanizing the rubber, and this probably resulted from the use of conventional proportions of curing agents. We now know that low molecular weight rubber requires a higher proportion of sulfur and vulcanization accelerators than does solid rubber.

C. Azel Prince, Jr., a chap of 18 years just out of high school, started working for Edison in July 1930 at West Orange (6). He was the son of C. A. Prince, a former druggist and Edison's chief chemist for the rubber project. The Prince family lived in Fort Myers. Edison moved certain of his key personnel, as well as certain equipment, back and forth from West Orange to Fort Myers as he changed his residence. A druggist might appear to be a questionable choice to head up such laboratory work, but considering that in the 1920's most drugs came from natural sources, one skilled in the chemistry and botany of drugs was really a very good choice.

Azel Prince, Jr., one of the few people still living who worked directly with Edison on the rubber project, now resides in St. Petersburg. He has kindly filled in many of the gaps of information for the year or so before Edison died and what happened to rubber from goldenrod after his death. Azel's job was to keep a daily record of plants tested as to laboratory number, name of plant, and percent rubber obtained. When the extraction technique was used, he saw that the sample of extract was placed on Edison's desk. When Edison was in the laboratory area, Prince was also responsible for having warm milk available at two-hour intervals. Warm milk and crackers were Edison's principal diet the last few years of his life. At least one cow was kept at his residence to provide the fresh raw milk.

Prince worked at West Orange until December 20, 1930 and then returned to Fort Myers. Edison and other members of his party came south January 1, 1931, and the work continued there. By early spring it was evident that Edison was a very sick man. He was transported in a wheel chair from the porch of his home across McGregor Boulevard to the rubber laboratory where he had a desk and chair. Tourists would often line the street near the Edison home, and some would attempt to take his picture as he was wheeled across. To avoid being so photographed, Edison would lean over and cover his face with his arms.

Edison would always give specific instructions as to what equipment to ship north well before moving day. In 1931 he did not leave Fort Myers until early June, which was one or two months later than he usually left. As the time of departure came near and still no word from Edison, young Prince was requested by the laboratory and field personnel to find out what Edison wanted shipped. Prince wrote a note to this effect (Edison was completely deaf at this time) and handed it to him one

Edison, in conference with two laboratory assistants at his Fort Myers laboratory, ponders the next step

Courtesy Edison Winter Home, Fort Myers, Fla.

morning while he was sitting on the porch. He got a one-word answer, "Everything." Edison knew that he would never return to Fort Myers.

During the summer of 1931 Edison went to the laboratory at West Orange only two or three times. However, he still requested reports on the work up to within a few weeks of his death on October 18, 1931. Early in October, the laboratory personnel made major improvements in the vulcanization of goldenrod rubber, and small samples were delivered to him. He appeared to delight in handling them, and observing their resiliency and snap. When Harvey Firestone visited him a few days before the end, Edison was too weak to speak, but his eyes indicated the pieces of vulcanized goldenrod with as much as to say, "Well, we did it."

Robert Halgrim, presently curator at the Edison Winter Home, is probably the only person now living in Fort Myers who worked with Edison. As a young man he worked in the botanical gardens and the rubber laboratory as time permitted. He worked on developing the "quick green test" using bromination to follow the change of rubber content of leaves of the goldenrod resulting from age, position on the stalk, and soil conditions.

Following Edison's death, work on the rubber project continued with J. V. Miller, Mrs. Edison's brother, in charge. With America in the depths of a severe economic depression, the price of crude rubber sank to 2⅝¢ in 1932. For a year and a half it lingered below 5¢. Firestone stopped shipments from his Liberia holdings. New plantings on the world's rubber plantations stopped. This was hardly the economic atmosphere to encourage work on rubber from goldenrod or any other alternate source. In September 1932 Harry G. Ukkelberg, fresh with a master's degree in agricultural science from the University of Minnesota, was hired to take charge of the work at Fort Myers. All subsequent work was restricted to various species of goldenrod; the botanical work was done at Fort Myers and the analytical testing at West Orange.

The author has had an opportunity to read a series of progress reports issued by Ukkelberg during the 1932–1935 period (7). Although the overall program appeared to have been poorly coordinated by Mr. Miller, a comprehensive program of considerable magnitude was carried out at Fort Myers. In 1933 a total of 288 experimental beds were planted containing 24,008 plants. In late summer and early fall, leaf samples were taken, dried, and shipped to West Orange for analyses. In 1933 a total of 6,664 such samples were shipped. Leaf samples were taken about mid-height of a given plant, usually at full bloom. From the analyses of individual samples and estimated production of dried leaves per acre, which was as high as 4,000 lbs, yields of rubber were calculated. In 1933 one plot showed a yield of 280 lbs of rubber per acre, and data for 15 experimental beds averaged over 200 lbs per acre. Rubber contents of leaves were as high as 12.4%. In 1934 when the harvesting practice for test samples was changed so that all the leaves from a plant were collected, rubber contents were still as high as 11.0%. In 1934 several calculated yields were over 300 lbs per acre. The 1934 results were based on analytical testing by the U.S. Department of Agriculture at Washington.

To one attempting to evaluate these data, the big unknown is the quality of the rubber and how much could have actually been isolated in a practical manner from an acre of goldenrod. The enormous number of samples were tested by bromination of the extract slurry although no rubber was actually isolated. On request of Harvey Firestone for a large sample for testing, a few hundred pounds of dried goldenrod leaves had been shipped to West Orange where they were extracted, and the rubber obtained was sent to Firestone. Ukkelberg never received any report on the testing by Firestone although he heard unofficially that the sample was high in resin content.

Courtesy H. G. Ukkelberg, Richmond Hill, Ga.

Harry Ukkelberg stands beside a rank growth of goldenrod at Fort Myers (1934).

In 1934 when the Edison family decided to close out the rubber project, arrangements were made to turn all planting material and records over to the U.S. Department of Agriculture. C. A. Prince and his son had gone to Washington earlier to discuss the matter with Dr. Polhamus, then head agriculturist of the Rubber Crops Section, USDA. Polhamus had continued to be quite interested in the overall project and had visited Edison at Fort Myers relative to his botanical studies dealing with goldenrod. He also carried out independent determinations of the rubber contents of 39 species of goldenrod and published these data in 1933 (8).

Polhamus came to West Orange in 1934 and with the help of a grant of $12,000 from Henry Ford prepared a card file covering all samples tested to that date. Considerably later (1967) this information was published in the form of a report by the Agricultural Research Service, USDA (5). This excellent compilation by Dr. Polhamus includes 56 different species of *Solidago* (goldenrod) tested. The number of samples of each species ran as high as 51.

During 1934 the program continued at Fort Myers with studies not only on crossbreeding and methods of propagation but also on such variables as soil fertility, amount of sunlight during the growing season, the degree of maturity when harvesting the leaves, and the like. Beginning in 1934 the analytical determinations on the leaves were done by the USDA in Washington. Cooperative work was carried out on a scab fungus and nematodes, both of which were becoming troublesome in the test beds at Fort Myers. In the fall of that year 22,400 goldenrod plants were shipped to the USDA Experimental Station near Savannah, Georgia. In March 1935 an additional 18,000 were shipped. Ukkelberg subsequently spent a week at Savannah demonstrating his methods of controlled crossing. This experimental station had been set up primarily for the purpose of developing crops suitable for the large tract of heavy gumbo-type soil which extends from South Carolina to northern Florida. Goldenrod, however, grows best in a sandy soil.

Henry Ford continued to have an interest in the goldenrod studies and, as mentioned above, financed the gathering and recording of available test data by Dr. Polhamus. After work was discontinued at Fort Myers by the Edison Botanic Research Corporation in 1936, Ford hired Ukkelberg to continue the goldenrod work on a limited basis on some land which he was then developing near Richmond Hill, Georgia. Ford had bought 80,000 acres along the Ogeechee River southwest of Savannah. Although most of this land was in timber, a few hundred acres were operated as a plantation primarily for experimental purposes. For example, 200 acres of tung trees were planted. Native to China, these trees produce a drying oil, commonly known as China wood oil, for use in paints. Ukkelberg shipped 10,000 goldenrod plants to Richmond Hill for his studies there and some work began. "In about a year, however, he (Ford) told me that synthetic rubber had been developed to the extent that he saw no need to continue the work and it was discontinued" (7).

Ford had continued to vacation at his winter home in Fort Myers for two years after Edison's death. After that he spent his winters in Georgia. He restored an old plantation mansion on the bank of the Ogeechee and had a workshop in a nearby building which had formerly housed a cotton gin. Ukkelberg continued to work for Ford after the goldenrod work at Richmond Hill was discontinued.

From 1934 to World War II, the U. S. Department of Agriculture continued a very modest program on goldenrod; their overall budget for work on rubber from plants was only about $20,000 annually.

Successes on the synthetic rubber front, which were beginning to be reported in the technical and public press, were a factor in greatly reducing all work worldwide on rubber from plants other than the *Hevea* tree. Intercontinental Rubber Company substantially reduced its production of guayule rubber. The Russians turned to synthetic rubber *via* sodium catalysis of butadiene. However, it was in Germany where most research and development on synthetic rubbers were done. Work in the 1920's was first concentrated on raw materials. Several cheap processes were developed for making butadiene, the simplest of the conjugated diolefins. Then a highly improved technique was developed for polymerizing the butadiene to rubber in emulsion to give a latex very similar in handling characteristics to that of natural rubber. On August 20, 1929 the I. G. Farbenindustrie Aktiengesellschaft, the principal German company conducting research in the field of synthetic rubber, filed a patent which also issued in the United States as Patent 1,938,730 on December 12, 1933. This epic patent and its equivalents in other countries covered as a new composition of matter a rubbery copolymer of butadiene and styrene. This elastomer, known as Buna S in Germany and subsequently under such names as GR-S and SBR in the States, was destined to become the major general-purpose synthetic rubber, not only during World War II, but also present day. In 1933 a 5-ton sample of German synthetic rubber was furnished to an American rubber company for testing. These major developments by the Germans as well as certain developments in the United States on specialty synthetic rubbers were well known in rubber circles.

It is beyond the purpose of this book to describe further the successful development of synthetic rubber in this country and abroad. Adequate accounts are available in technical encyclopedias such as the one cited (9). A more detailed account of the development in the United States prior to and during World War II is given by Frank A. Howard, a leader in making possible the availability of German synthetic rubber know-how to American scientists and engineers (10).

As Edison had predicted, rubber became a critical item in World War II. Although the worst that had been expected was that it might be difficult to keep the sea lanes open to the

Far East, the capture of Malaya, Singapore, and the Dutch East Indies by the Japanese in late 1941 and early 1942 was catastrophic. The only plantation rubber left in possession of the Allies was that in Ceylon and Firestone's holdings in Liberia. Little has been written about what was done during the war years to make natural rubber available from alternate sources, including goldenrod. Thus, this will be described in some detail since it includes the final work carried out on rubber from goldenrod.

Although a vigorous and well-directed program on synthetic rubber was proceeding in early 1942 under the auspices of the Rubber Reserve Company, Department of Commerce, much confusion and rivalry existed on the issue in political circles and among certain segments of industry as to who was best suited and qualified to produce the various raw materials and the synthetic rubbers. In fact, the question whether butadiene should be prepared from petroleum or *via* ethyl alcohol produced by the fermentation of grains became a national issue. Congress jumped into the act and passed a rubber supply bill on July 21, 1942 providing the establishment of a new agency to make butadiene and rubber *via* alcohol from the fermentation of farm products. This legislation is ample evidence that we need Congressmen who are technically trained as well as lawyers and businessmen. Technical information was available at that time to the effect that it was cheaper and would require less strategic materials to manufacture butadiene from the C_4 gases from the cracking of petroleum than from alcohol. The record shows that alcohol was only used to make butadiene during an interim period at one site. President Roosevelt properly vetoed the act of July 21 and in an attempt to create a national rubber policy which would have the backing of all, requested Bernard Baruch to head up a committee to evaluate the rubber situation.

This illustrious committee also had as members James Conant

and Karl Compton. The former was then the president of Harvard University and a famous organic chemist. Compton was president of the Massachusetts Institute of Technology and a former professor of physics. The committee reported its findings on September 10, 1942. Although this report contained little in the way of new information and recommendations over that of the Dickinson Committee of June 25, 1940 (11), it did serve to unite the country behind the rubber program and to establish high priorities for men and materials to expedite it. The manufacture of the various synthetic rubbers was to carry the major share of the load for providing rubber for both the war effort and civilian uses. There were other major recommendations. These included a nationwide speed limit for autos and trucks of 35 miles per hour to decrease tire wear per mile, rationing of gasoline to curtail unnecessary driving, and the recapping of used tires with reclaimed rubber to be collected by a major drive to have the public turn in worn-out tires and other rubber articles.

As for natural rubber from plants in the western hemisphere, the Baruch Report cited the following possibilities:
(1) Goldenrod, based on Edison's experiments
(2) *Kok-saghyz,* a dandelion studied by the Russians
(3) Guayule
(4) *Cryptostegia*
(5) Wild rubber trees growing in South and Central America
The Baruch Committee recommended work only with the last three cited possibilities. The wild rubber industry, which was largely in Brazil, had been in a dying state the previous 25 years, so it was hard to revive. Whereas Brazil rubber had been selling on the New York market for 20¢ per lb before Pearl Harbor, the price was pegged at 45¢ in 1942 to encourage production. In February 1944 the price was raised to 60¢. Skilled tappers from the era up to about 1920 had died or moved to other areas. Over 18,000 laborers were moved into the Ama-

Rubber from Goldenrod

zon valley from south Brazil by August 1944 to collect rubber latex from the wild *Hevea, Castilla, Manicoba,* and *Mangabeira* trees. The amount of rubber shipped from Brazil's jungles proved to be a grave disappointment to some, but to those familiar with the growing of wild rubber, the amount of rubber obtained was up to expectations (*12*). The Committee probably overlooked the fact that Latin America needed its own rubber during the crisis, as indicated by the fact that 20,000 tons of Far East rubber had been imported in 1939.

Although the Intercontinental Rubber Company continued some activity with guayule in the United States during 1925–1939, it was at a low level. The very low prices for crude rubber made production of guayule rubber highly uneconomical. Farmers in California had contracted to grow guayule for Intercontinental, but much of this acreage was plowed under before the crop was harvested. The extraction mill built at Salinas in 1930 was operated four times for brief intervals during the next 10 years.

In 1930 the U.S. Army carried out a study of the guayule industry for the Quartermaster Corps. One of the two officers making the survey was Major Dwight Eisenhower. Their report included the recommendation "that the Government develop guayule as insurance against the complete absence of a rubber supply in event of war." A survey by Intercontinental Rubber indicated that 5,600,000 acres of land in southwestern United States were suitable for growing guayule. Of this, 2,500,000 acres were range land, containing largely brush. Representative Anderson of California pleaded for government assistance in developing rubber from guayule. However, it was not until after Pearl Harbor that Congress acted. On March 5, 1942 the President signed Public Law No. 473 of the 77th Congress authorizing the Secretary of Agriculture "to provide for the planting of guayule and other rubber-bearing plants and to make available a source of crude rubber for emergency and defense uses." The original bill provided for the planting of 75,000 acres of guayule, but following the Baruch Committee Report, the bill was amended to increase the acreage to 500,000. The Secretary of Agriculture designated the Forest Service as the Departmental agency to administer the program (13).

The California guayule properties of the Intercontinental Rubber Company were acquired by the Government. The purchase included 1,483 acres of land in the vicinity of Salinas, which is just inland from the Monterey Peninsula. Included also were a factory and several guayule nurseries. Plantings started immediately, and by April 26 about 9,935,000 seedlings had been planted on 808 acres as well as 70 acres of "indicator plots" distributed in various parts of California, Arizona, New Mexico, and Texas.

In January 1943, Rubber Director Jeffers directed that the program on guayule be curtailed since it appeared that the production from additional seeding, which would not come in until 1946, would be of doubtful value in the war effort. This recommendation was confirmed in the Rubber Director's Progress Report of May 3 in which he commented, "Conditions have changed. Food for 1943 has become more important to the nation than a rubber insurance policy for 1946 and beyond." Proposed acreage for guayule in California had included irrigated land, which was highly productive for food crops. Thus, in spite of the proposed planting of 500,000 acres to guayule, only 32,000 acres of plantations were established.

In the spring of 1943 guayule from a 550-acre field in California planted in 1930, and subsequently acquired by the Government, was processed to give 393 long tons of rubber. In 1943–44 the native guayule shrubs growing wild in the Big Bend Area of Texas were gathered and processed to produce 225 tons. Thus, a total of 618 long tons of rubber was produced during the war years from domestic guayule. The total cost of the guayule program, which received its last appropriation in the July 1944–July 1945 fiscal year, was $37,700,000.

Rubber from the 1942–43 plantings was never harvested. The acreage was plowed up for other crops except small areas retained by the Department of Agriculture for experimental purposes. Studies with guayule continued in both California and Texas until 1955. The last Department of Agriculture appropriation which included any money for work with guayule was for the fiscal year 1954 and for only $108,000.

Edison's search for rubber from plants is believed to have inspired similar work in Russia. An article entitled "Natural Rubber from Russian Dandelion" published in January 1946 (*14*) stated:

In the course of a systematic investigation instituted in Russia in 1929 to determine the possibilities of producing natural rubber in that country, the *kok-saghyz* was found near Tien Shan, Kazakstan, near the border of China. By 1932 more than 2,000 acres of this plant were under cultivation in Russia, and it is reported that at the time of the German invasion of Russia approximately 200,000 acres of *kok-saghyz* had been planted for the production of rubber.

Shortly after the tragedy at Pearl Harbor the United States Government started negotiations with Russia to obtain *kok-saghyz* seed. These eventuated in the arrival in Washington on May 8, 1942, of two gunny sacks of seed flown from Kuibyshev. On the same day the seed arrived some of it was repackaged and distributed by air to various points in the United States, where more than 60 indicator plantings were established through cooperation with state agricultural experiment stations.

The Emergency Rubber Project authorized by Congress as mentioned above included authorization to work on "other rubber bearing plants" besides guayule. Although the Forest Service of the USDA was responsible for the work, they were free to call upon other bureaus of the Department for assistance. In the 1930's four regional research laboratories had been set up by the Agriculture Department to seek outlets for farm crops as raw materials for producing industrial products. Thus, based on personnel and equipment available, these laboratories were excellently qualified to carry out and evaluate the recovery of rubber from plants.

The Eastern Regional Research Laboratory, located at Philadelphia, Pennsylvania, was given the responsibility not only for developing improved methods for the recovery of rubber from guayule but also that of investigating the potential of the Russian dandelion, *Taraxacum kok-saghyz*. Plants grown from the seed furnished by the Russians were shipped to the Philadelphia Laboratory in the fall of 1942. Analyses showed that more than 90% of the rubber is in the form of latex in interconnecting root ducts. The remainder of the plant does not contain sufficient rubber to justify its recovery. A process was worked out in which the roots were field or oven-dried to 15% moisture which served to coagulate the rubber; they were then water-leached at 200°–212°F to remove soluble carbohydrates, and the rubber was recovered by a sequence of pebble milling, screening, flotation, and drying.

Yield of roots was about 500 lbs per acre on a dry weight basis containing about 12% rubber, which amounted to a low yield of about 60 lbs of rubber per acre. Based on a limited study by the Agriculture Department, the best growing area was found to be in the Red River valley of Minnesota and certain sections in Michigan. Crops would be planted in early spring, and the roots would be harvested in late fall or the following spring. The roots also contain 10% or more of carbohydrates which appeared to be suitable for fermentation to ethyl alcohol or lactic acid.

The rubber isolated from these studies was of excellent quality having a molecular weight in the same range as that of *Hevea* rubber. Although it was only 80% pure as recovered, the impurities were largely finely divided inert material which did not interfere in subsequent mixing, fabricating, or vulcanizing.

310

Experimental passenger car and truck tires fabricated by the B. F. Goodrich and the United States Rubber Companies indicated that *kok-saghyz* rubber was equivalent to *Hevea*.

Work on *kok-saghyz* in the United States was terminated on June 30, 1944. Although this plant gives a high quality rubber, the necessity for using fertile land to produce it, the relatively low yield of rubber per acre, and the high cost of harvesting a root crop has apparently resulted in no extensive studies with the Russian dandelion since World War II.

The Baruch Committee felt that *Cryptostegia,* a perennial vine of the milkweed family, growing in Florida, Mexico, Haiti, and India, was second to guayule as the best possible alternate source of natural rubber. This recommendation was definitely in error. It was true that considerable information was known about rubber from *Cryptostegia*. It had been a favorite plant for studying the synthesis of rubber in nature since rubber occurs both as a latex in a duct system in the stems and extending into the leaves paralleling the vein system, as well as occurring in leaf chlorenchyma cells. Edison had studied *Cryptostegia* in his early rubber studies in 1927 and had rejected it as a possible rubber source.

Activities relative to *Cryptostegia* outside the United States were handled by the Rubber Development Corporation which had been set up by the Government under the Foreign Economic Administration to facilitate recovery of rubber from Latin America. In 1942 this Agency contracted with a Haitian Agricultural Corporation to establish 100,000 acres of *Cryptostegia* in Haiti. *Cryptostegia* had been introduced as a garden ornamental in Haiti in 1908 and had become well established over a considerable area of the Gonaives Plain. An integral part of the program was a research laboratory at Gonaives, Haiti. Extensive experimental studies were carried out on how best to collect the latex from the vine (*15*). This proved to be much more difficult than recovery from the *Hevea* tree and is probably impractical from an industrial point of view. The rubber from the latex exuded from the living *Cryptostegia* is of high quality although yields are low.

The *Cryptostegia* vine had never been cultivated commercially. Of the 100,000 acres planned, only 43,000 were actually planted although an area of 70,750 acres of land leased from the nationals was cleared. It was found that the plant was quite selective to soil and moisture conditions, requiring much longer to mature than had been expected. It became apparent by April 1944 that no rubber could be obtained from this source in 1944 or 1945. This resulted in the Rubber Development Corporation's terminating the project and turning the land back to the nationals. Some experimental work continued at the Gonaives Laboratory. The loss on the overall project was $6,343,300. A large-scale *Cryptostegia* plantation project in Mexico was discontinued before any substantial investment had been made.

In the 1930's Dr. Polhamus then at the Bureau of Plant Industry, USDA, continued Edison's work on goldenrod as limited funds permitted. Polhamus encouraged the Forest Service to include goldenrod in its studies "of other rubber-bearing plants" as authorized in the Emergency Rubber Project. The Southern Regional Research Laboratory located at New Orleans was authorized to evaluate goldenrod as a possible source of rubber during the war emergency. The work began in June 1942, and at times there were as many as 20 professional people assigned to the project (*16*). Fifty acres of goldenrod consisting of selected strains of the *leavenworthii* species were already planted near Waynesboro, Georgia, a few miles south of Augusta. Work in 1942 was with almost pure leaf material hand-stripped from this 50-acre growth.

A comprehensive study was initiated by the Southern Regional Laboratory. One group carried out microscrope studies on goldenrod leaves (*17*). Stains were developed which would selectively color the rubber and resin globules and thus make possible meaningful color microphotographs. The so-called resin proved to be a very complex oil mixture containing among other things, 12% sugars, 11% of quercetin, a polyhydroxy derivative of a polyheterocyclic, and organic acids and saponins, which are condensates of steroids and sugars. Some constituents of the resin can be extracted from the leaves by hot water and are used to make a tea and to produce an oil used as a tonic; on the basis of these products one species is known as sweet goldenrod. The rubber globules were found to be in cells separate from those of the resin. The rubber globules are about 2 microns (0.000079 inch) in diameter, and as many as five were found in a single cell. Whether the rubber in this finely divided form is in solid or liquid state could not be determined.

Laboratory studies with leaves from the 1942 crop indicated that the best recovery technique was liquid extraction with acetone to remove the bulk of the resin, followed by benzene to dissolve the rubber. This is quite different from the acetone–benzene Soxhlet extraction where the sample is held in a porous cup below a reflux condenser. The pilot plant consisted of eight cells, each 12 inches in diameter and 5 feet high and holding 75 lbs of ground leaf material. The extraction equipment was available from previous studies on extracting rosin from pine chips. Acetone at 140°F was first pumped through the cells for about six hours. This removed 13–18% of the material. After the acetone was drained off, benzene at 175°F was passed through for nine hours. The benzene contained 0.5%, based on the rubber, of Flectol H, a common rubber antioxidant. The benzene was concentrated to 20% rubber by evaporation, and acetone was added to precipitate the rubber. This served to decrease resin in the rubber from about 25 to 5%.

All solvents were recovered and recycled. Total operating costs including loss of solvents indicated a cost of 16.4¢ per lb of crude rubber recovered. This did not include the cost of growing and harvesting the goldenrod. Also, no credit was allowed for the by-product resin obtained as the acetone extract.

Unfortunately, the 1943 crop of goldenrod was not harvested until blooming had progressed well, and it contained only 20–25% leaf material. An attempt was made to remove the non-leaf material by screening, but still the rubber content of the goldenrod used in the extraction studies was only 2.7%. A rice binder was used primarily for cutting and recovering the goldenrod.

In the extraction studies, the degree of fineness obtained on grinding the leaves was found to be a highly important variable. Complete rubber removal was obtained with leaves ground 90% through 40 mesh compared with partial removal with 30-mesh powder. Extensive pilot-plant studies were carried out in the fall and winter of 1943 using the 2.7% leaf material. A total of 874 lbs of rubber, calculated on a resin-free basis, was accumulated from this work.

The Southern Regional group confirmed Edison's work that goldenrod rubber is a low molecular weight sticky material. Although it is identical to *Hevea* rubber chemically and in the structural arrangement of the atoms within the molecule, the number of combined isoprene molecules in a rubber molecule is much less in goldenrod rubber. As a final step in isolating the goldenrod rubber, it was mixed with sulfur and various organic accelerators in benzene solution, and the solvent was removed under 14 inches of vacuum while mixed in a kneader. The temperature was then increased to 250°–280°F to vulcanize the rubber partially, thus converting it to a non-tacky material (*18*). The rubber was then ready for mixing using conventional rubber equipment. Tensile strength values as high as 3,000 psi were obtained with the vulcanized precured rubber, whereas 1,300 psi was maximum when starting with the raw goldenrod rubber. The precured rubber was quite stable, as indicated by storage tests of over a year.

Both raw and precured goldenrod rubber were distributed to several rubber manufacturing companies for fabricating rubber articles using factory equipment. Results of these tests include the following (*16*):

Raw goldenrod rubber, as prepared by procedure previously given, can be blended with GR-S synthetic rubber up to at least 30% of goldenrod rubber with desirable results, and that blending increases building tack, reduces absolute hysteresis, and increases elongation of the product. In addition goldenrod rubber can replace up to 20% of *Hevea* with desirable results.

Precured goldenrod rubber has desirable properties and can be satisfactorily processed by milling, calendering, and extruding in the form of tread stock with carbon black filler.

Various commercial objects, including heels for shoes, hot water bottles, small sample tires for use with ash trays, and rubber-impregnated belts were prepared on factory equipment for the primary purpose of testing the processing practicability of precured goldenrod rubber. It was found that these objects could be made using factory equipment.

The relatively high tensile strength, low hysteresis, good flex resistance, and excellent bonding of goldenrod rubber to fabric were reflected in the mileages indicated on wheel tests of three bicycle tires. These mileages, besides being three times that of reclaim casings then produced under war conditions, were nearly double those of prewar crude rubber casings. Several of the tires manufactured from goldenrod rubber were issued to private individuals for use on their own bicycles. The mileage on one set of these tires is reported already to be approximately 1,500. All of the tires, after two years of running under actual service conditions, are still in use.

Work on rubber from goldenrod was terminated in June 1944, as were the other projects which came under the Emergency Rubber Act of 1942. The outstanding success with synthetic rubber caused the demise of the goldenrod project rather than lack of promising results from the latter. It is believed that no subsequent work has been done on rubber from goldenrod since World War II except the compilation of Edison's findings on plants (5).

Mr. Ukkelberg has commented relative to why Edison's work on goldenrod as a source of rubber failed to reach industrial importance (7):

One of the big reasons was that he did not have time enough. Plant improvement is a slow process. You have to wait a whole year before results from any test or trial are available. Even today with all the knowledge in genetics and plant breeding that has been acquired in the past fifty years, it takes from six to ten years to develop an improved variety of almost any crop and then you cannot be sure it will be a commercial success. The science of genetics and plant breeding was comparatively new in Edison's time and few people understood the principles involved except those who had special training. I found no evidence that Mr. Edison had any information in this field nor did he have anyone with him who did. The fact that he was making selections within colonal lines (a colonal line being the progeny of a single plant propagated vegetatively) would indicate that. One of the principles of plant breeding is that selection within a colonal line is futile except in the case of a rare mutation which may or may not be an improvement, usually not. He started from scratch, you might say, with a plant about which practically nothing was known and in a field in which he had very little information. That, of course, was nothing new for him but it would take time for anyone, including Mr. Edison, to acquire this information.

Much more is known today about handling liquid and semi-solid elastomers than was known in 1944. In fact, the synthetic rubber industry has tried to develop liquid rubbers which can be vulcanized to products comparable with those of solid rubbers. Liquids are cheaper to mix with fillers and other additives than are solids and require less energy and less rugged equipment to process and mold them. In fact, a recent (1970) announcement by the Firestone Tire & Rubber Company disclosed "a revolutionary process for tire manufacturing, consisting of injecting hot, liquid rubber into a mold and minutes later removing a cured, fabric-free tire" (*19*). Furthermore, based on extensive experience of certain manufacturers of synthesizing elastomers in solution, the recovery of a rubber from a solvent such as benzene is now an accepted routine factory operation. Thus, low molecular weight natural rubber from goldenrod might have a premium value in contrast to *Hevea* rubber of similar hydrocarbon content.

Harry Ukkelberg had apparently made a good start in 1934 when, under highly favorable growing conditions, he obtained results indicating yields up to 300 lbs of rubber per acre of goldenrod. Our present increased knowledge on the effective use of fertilizers should be helpful in increasing yields as well as in improving the plants. Goldenrod rivals corn in total weight of dry product which can be produced annually. Its leaves contain more resin than rubber, and even its non-leaf portion may have value. Goldenrod is a perennial and can easily be propagated by rootstock cuttings. Does it offer today a possible means of putting to worthwhile production idle acres in our southern states?

Rubber from guayule received some Government support for research and development for about 10 years after World War II. The reason for this was that in spite of adequate capacity for synthetic rubber during an emergency, the industry still needed some natural rubber to make an acceptable truck tire. However, in the 1950's another major advance was made in the field of synthetic rubber. This was the highly important development of a catalyst system by Karl Ziegler of Germany whereby olefinic hydrocarbons could be converted to polymers of controlled structures. Thus, for the first time natural rubber could be duplicated synthetically. Two U.S. companies began manufacturing *cis*-polyisoprene in the late 1950's. The rubber plantations in the Far East continue to make major advances in increasing the yield of rubber from *Hevea*, and natural rubber remains competitive with synthetic. Nature's share of the total rubber market worldwide for 1970 was 38.8%. The wild fluctuations in the price of natural is a thing of the past, and a ceiling of about 30¢ per lb will probably exist for many years to come.

Some have said that Edison missed the boat when he undertook work on rubber from domestic plants instead of working on synthetic. If so, he was not alone. One of the papers given before the symposium on raw rubbers sponsored by the Rubber Division of the American Chemical Society in 1926 was entitled "Future Commercial Prospects for Synthetic Rubber," by William C. Geer (4). Mr. Geer was a former vice-president of The B. F. Goodrich Company, one of the so-called Big Four rubber companies of that era. The final conclusion of his paper was as follows:

So the world has little to expect, and the planters nothing to fear, from synthetic rubber, and in the inevitable cost competition which would arise were synthetic rubber to be produced, the planter would be able to deliver crude rubber on board ship at a profit to himself, and at a price to which synthetic rubber could not be brought.

Although the Americans developed certain specialty rubbers and manufactured them in a limited way prior to World War II, it was the Germans in their drive for self-sufficiency who developed general purpose synthetic rubber and demonstrated its use in tires. The Standard Oil Company (New Jersey) held the patent rights of the basic German inventions on synthetic rubber for the United States but did not consider general purpose synthetic rubber a promising business venture. It chose to license the patents to the rubber companies, who in turn felt that Government subsidies were necessary. Thus, it took the rubber crisis of 1941–45 to force America to build a synthetic rubber industry, which quickly became competitive with nature's product.

Edison like many others did not foresee the potential of synthetic rubber, but he did foresee that the United States should provide an alternate source of rubber in case war shut off our overseas supplies; moreover, he attempted to do something about it. He must have been very concerned since he was 80 years of age, and he knew that success would not directly benefit himself or his personal business. Although his work on rubber did not help us during World War II, our rubber supply situation in 1942 would have been less critical if more leaders in industry and Government had been as concerned as was Thomas Edison.

The present era has been described by some scientists and engineers as the hydrocarbon age. Today petroleum and natural gas supply 76% of the total energy used in the United States, with coal, hydroelectric, and atomic energy supplying the other 24%. Whereas before World War I most organic chemicals were prepared from coal, products of the forests, and farm crops, such chemicals are now synthesized largely from petroleum and natural gas. Most synthetic elastomers and many plastics are derived wholly from these natural hydrocarbons. Even proteins for use in animal feeds are being prepared from petroleum on a small scale. However, fossil hydrocarbons are obviously limited in supply, and some day, possibly centuries in the future, man will turn again to the soil as the primary source of materials. Then someone will likely become famous by rediscovering that goldenrod is a potential source of rubber.

Chapter 10 **Edison's Place in the Chemical Fraternity**

THOMAS EDISON is usually thought of as an inventor. This he certainly was; he was granted 1,093 U.S. patents, more than any other person. Usually an inventor is a scientist or an engineer. The present-day inventor tends to be more specialized than these two categories suggest; he is a particular kind of scientist or engineer. Assume, for example, he is a chemist. The next question is, what field of chemistry? There are literally a dozen categories in this classification. Edison, however, was not a specialist. When asked what his field of interest was, he once replied, "Everything."

As Robert Millikan has pointed out, Edison lived in an age when the average person was ill-housed, ill-clothed, and ill-fed (1):

In the United States the period into which Mr. Edison was born was one which called loudly for the increase in material facilities. It was a new and a largely undeveloped country calling violently for transportation, for communications, for the means to make one man's labor produce the maximum of goods. Edison heard that call, saw that need, and bent his matchless energies and capacities to meet it.

The big opportunity in life, as Edison saw it, was to develop useful things that people would want to buy because they gave them a better life. The mechanism of doing this was to utilize the Patent Office to get the 17-year protection which allows an inventor time to bring his new development to fruition and hopefully to recover his costs and make a profit. Although Edison often complained of the shortcomings of the Patent Office, he found that getting patents was a part of the process of developing new things.

He lived in an age when the reservoirs of scientific knowledge were sufficiently low that an inventor could acquire a working knowledge in several fields. He was doing chemical work before the American Chemical Society (ACS) was born and had been active in applied science for 40 years before the Society's first journal in applied chemistry became available. Founded in 1876, the ACS consisted primarily of academic and government agency chemists. During its first 32 years, only two presidents of the Society could be called industrial chemists. One was Charles Dudley, chief chemist for the Pennsylvania Railroad; the other was James Booth, founder of an analytical laboratory in Philadelphia. Neither could be considered a research man. George Barker, professor of physics at the University of Pennsylvania was president in 1891. Outstanding chemists in America were obviously few and far between, and the future didn't look promising. Harvey Wiley, chief chemist of the U.S. Department of Agriculture and "father of the pure food law," predicted in an address before a meeting celebrating the 25th anniversary of the Society in 1901, that its membership in the centenary year, 1976, "will be nearly 10,000." Actually, membership reached ten times that prediction in 1965.

Edison had carried out his most important work by the time he was 70 years of age. Most chemists have retired before then or are busy talking or writing about their experiences. Taking this arbitrary productive limit in Edison's case brings us to the approximate end of World War I. That is a good dividing line to consider since there was a radical change in American chemistry in the post-war period. How do Edison's contributions to applied chemistry compare with other American inventors prior to World War I? The following six would appear to have been Edison's greatest contemporaries in the field of overall applied science and in applied chemistry in particular.

George Westinghouse (1846–1914) served in both the Union Army and Navy during the Civil War. Subsequently he attended Union College but left after three months to work in his father's shop in Schenectady. At 19 he invented a rotary steam engine and a device for replacing derailed cars. In 1869 he invented the air brake for railroad cars. The practice prior to that time was for a brakeman on each car to apply the brakes on a signal from the engineer. Westinghouse's invention of the automatic air brake in 1872 won him world renown. He also developed a system of railway signals activated by compressed air and electrical devices.

The Westinghouse Electric Company was organized in 1886, and in 1887 it acquired several small electric companies including the Consolidated Electric Light Company, which controlled the Sawyer-Man patents. George Westinghouse's first big test in the electrical field came at the time of the Columbian Exposition of 1893 in Chicago. He had won the contract for lighting the Exposition and had planned to use the Edison-type electric lamp which Westinghouse and others were manufacturing regardless of the patent situation. In 1892 following a long litigation, Edison's U.S. Patent 223,898 was sustained by the courts. Westinghouse now was no longer free to make the sealed-glass–carbon filament light. To circumvent the Edison patent, Westinghouse developed a two-piece lamp using a ground glass joint at the base sealed with a cement. By superhuman effort the Westinghouse Company manufactured several thousands of such lamps and had them in place when the Exposition opened. Although the bulbs were evacuated, their life was considerably shorter than the one-piece bulb.

Courtesy Westinghouse
Electric Corp.

George Westinghouse

The Westinghouse exhibit at the Exposition demonstrated a 60-cycle polyphase alternating current of 2,200 volts reduced to less than 100 volts by transformers for powering the lamps. The Westinghouse Electrical and Manufacturing Company had championed alternating current whereas Edison had vigorously opposed it. Edison was not alone; *Scientific American* and Lord Kelvin also opposed the use of high voltage alternating currents until after Westinghouse's demonstration in Chicago in 1893. The first major generation of alternating current was at Niagara Falls in 1895. Westinghouse installed three 5,000-horsepower generators. Current was transported to nearby Buffalo at voltages as high as 22,000 volts. Westinghouse had won "the battle of the currents."

Westinghouse certainly was a major contributor to applied science in the 1870–World War I era. Edison's greatest accomplishment in the electrical field was the overall development of producing, distributing, and using direct current; that of Westinghouse, alternating current. Westinghouse was granted a total of 361 U.S. patents. Although he became most famous because of his air brake—a mechanical invention—his various developments in electric lighting, refinements in the handling and use of alternating currents, and his work in the field of copper refining utilized much applied chemistry.

Alexander G. Bell (1847–1922), one of the three sons of Alexander Melville Bell, was born in Edinburgh, Scotland. His father originated the phonetic "visible speech" system, commonly known as lip reading, for teaching the deaf. After a good education in London where the family was living, at the age of 20 Alexander became an assistant to his father. When tuberculosis took the lives of the other two sons and threatened that of young Alexander, the family moved to Canada and settled near Brantford, Ontario.

Young Bell got a job in Boston teaching the deaf. There he met Gardiner Hubbard, a wealthy lawyer who later founded and became first president of the National Geographic Society. Bell married Hubbard's daughter who had been deaf from age five following a severe attack of scarlet fever. Hubbard supported Bell's experiments in teaching the deaf and subsequently those on the multiple harmonic telegraph and the telephone.

Bell's telephone transmitter consisted of a strip of iron attached to a membrane which, when actuated by the voice, would vibrate in front of an electromagnet, thus introducing an undulatory electric current capable of transmitting speech. At the receiving end, a similar device was used in reverse. An electric battery activated the electromagnet. Bell filed his telephone patent application February 14, 1876. Edison greatly increased the distance of telephone transmission by use of the variable-contact carbon transmitter, the principle of which is still being used today. Bell did little on the telephone following his original invention.

Bell actually received little compensation from the Bell Telephone Company which was formed to exploit his patents. As mentioned in Chapter 4, he received the Volta Prize of 50,000 francs which he used to form the Volta Laboratory in Washing-

Courtesy American Telephone
and Telegraph Co.

Alexander Graham Bell

ton, D.C. Bell's share of the Graphophone patents obtained by the Volta Laboratory was $200,000. This windfall plus help from his wife's family served to finance his subsequent research. His future studies dealt with heredity of deafness, longevity, and aviation. Much of his later research was done at his summer home in Cape Breton Island, Nova Scotia. Bell contributed much to the beginnings of the National Geographic Society and of *Science* magazine, successor to the weekly by the same name published by Edison from 1880 to March 1882.

Edward G. Acheson (1856–1931) quit school at 17 to work in the coal and iron mines in the Pennsylvania area. Shortly thereafter he filed a caveat in the Patent Office dealing with an improved method for mining coal. He went to New York in 1880 in search of a job in the new electrical industry. He got one as a draftsman with Edison. Acheson worked at night on some of his own ideas when he could use tools in the laboratory shop. He frequently suggested new designs to Edison, most of which proved to be old or no good. One day Edison told him, "I do not pay you to make suggestions to me; how do you know but I already had this idea, and now if I use it you will think I took it from you."

Edison wanted Acheson to take over the lamp factory early in 1881 since Batchelor was scheduled to go to Paris in the spring. Acheson, who was making $7.50 per week, demanded $100 per month to run the lamp factory, and when Batchelor refused to grant it, Acheson left Edison's employ.

Unable to find other work, he returned to Menlo Park in a few days to see Edison. Edison laughed and joked about his not being able to take the hard work in the lamp factory and rehired him for experimental work.

Early the following summer, 1881, Edison sent Acheson to Paris to assist Batchelor at the Edison exhibit of the Paris Exposition. After this international electrical exhibit closed, Acheson stayed on with the French Edison Company at $150 per month. He left the French Company against the wishes of his superiors to join the Edison Italian affiliate at a higher salary. Finding his position there untenable, he quit and ended up in London sick and penniless. An Edison employee, James Holloway, whom Acheson had known at Menlo Park, took him in and got medical help. After recovering and learning that Edison's secretary, Samuel Insull, was in London, Acheson visited him at his hotel and related his circumstances. Insull cabled Edison who sent money for Acheson to return to America in January 1884. Edison wanted Acheson to become chief engineer at a new central light station being built in upper New York City, but he refused, saying he preferred experimental work. Thus, for the third time Acheson left the employ of an Edison company.

Acheson then worked for the Consolidated Electric Light Company, which was trying to duplicate the Edison light. He subsequently became general superintendent but left in late 1884 to join his brother who was attempting to reduce iron ore with natural gas. Shortly thereafter Acheson developed an anti-induction telephone cable which he sold to George Westinghouse for $7,000.

Courtesy Acheson
Industries, Inc.

Edward Goodrich Acheson

Henceforth Acheson's fortunes rose rapidly. In 1891 he synthesized silicon carbide by chance while experimenting with clays in an electric furnace. He formed the Carborundum Company and began manufacturing the new abrasive under that trademark name. Its first use was substitution for diamond powder in polishing gems. Acheson used Carborundum with a binder to make grinding wheels. His first big order was for 60,000 from George Westinghouse to be used in making the ground glass joints of the two-piece lamps for the Chicago Exposition.

In 1896 Acheson prepared artificial graphite, again utilizing the electric furnace. This invention led to many new products, including graphite electrodes for high temperature electrolysis and graphite emulsions as lubricants. Acheson promoted his two outstanding inventions with great vigor and became a highly respected and successful business man. Acheson, as Edison, certainly can be characterized as having been a very hard worker and an ardent experimenter (2).

Arthur D. Little (1863–1935) was a dropout from the Massachusetts Institute of Technology, yet later he became president of its alumni association. In 1884 he left his studies at M.I.T. to become chemist for the Richmond Paper Company, Rumford, Rhode Island, the first sulfite wood pulp mill in the United States. With Roger Griffin in 1886, Little set up a consulting and analytical service laboratory in Boston. Griffin lost his life in a laboratory accident in 1893. Little was then joined by William Walker, who later became famous as a professor of chemical engineering at M.I.T. A. D. Little, Incorporated pioneered the role of the industrial consulting laboratory in the United States and continues to be a highly successful organization.

At one time Little's consulting service had 60 papermakers as clients. He obtained basic patents on the preparation of cellulose acetate and licensed these to Eastman Kodak for film

Arthur D. Little

and to the Lustron Company for fibers. Other developments included processes for chrome tanned leather, the recovery of turpentine and rosin from pine stumps, paper from bagasse, as well as early work on vapor-phase cracking of petroleum. Perhaps his major accomplishment was demonstrating the value of research to the American chemical industry. Little published and lectured widely. He was president of the American Chemical Society in 1912–13.

Leo H. Baekeland (1863–1944), a Belgian by birth, received his Ph.D. in chemistry at the University of Ghent. Recipient of a traveling scholarship, he toured England looking unsuccessfully for university-sponsored research in applied chemistry. On coming to the United States, Baekeland was impressed by the work being done by Professor Charles Chandler of Columbia University in sanitation and other practical fields. Baekeland decided to remain in the United States and got a job with a photographic firm, the predecessor of Ansco. He later formed the Nepera Chemical Company and invented Velox, a photo printing paper with which artificial light could be used instead of sunlight, as had been the practice. He developed Velox to a practical reality and in 1899 sold his patents and know-how on the process to George Eastman on very liberal terms.

Baekeland then decided to take up electrochemistry and studied in Germany one winter on the subject. He made major

Leo Hendrik Baekeland

improvements in the Townsend Electrolytic cell for producing sodium hydroxide and chlorine from brine utilizing the "Baekeland Diaphragm." The outcome of this work was the formation of the Hooker-Electro-Chemical Company, Niagara Falls, in 1903.

Starting in 1905 Baekeland began work on the phenol–aldehyde reaction. His contributions to this field are related in Chapter 8. Baekeland received many awards in the chemical field and was president of the American Chemical Society in 1924.

Henry Ford (1863–1947) was largely responsible for the various mechanical innovations which resulted in automobiles so reliable that one not skilled in automotive mechanics could still drive with confidence. His production methods and the philosophy of a cheap car for the masses led to his colossal success as a business man, particularly during the 1908–1926 period. Ford pioneered research and engineering laboratories in the motor industry. Much of the early work carried out in these laboratories was of a chemical nature.

Ford obtained 161 U.S. patents over a period of 52 years—from 1898 to 1950. He worked on metal alloys and pioneered automation in a rubber tire plant and an improved process for manufacturing plate glass. Ford was a leading advocate for using farm crops as raw materials for making industrial products. His early work on soybeans was a major factor in making them the major farm crop which they are today. His experimental plantation at Richmond Hill, Georgia, and his rubber plantations in Brazil were valiant efforts to grow products from the soil as competitive raw materials for industry.

In addition to these eminent inventors who made major contributions in the field of applied chemistry, the following were also outstanding in this field during the Edison era.

James C. Booth (1810–1888). Founder in 1836 of a laboratory in Philadelphia dealing with analytical and applied chemistry; for 40 years melter and refiner at U.S. Mint in Philadelphia; author of then well-known "Encyclopedia of Chemistry."

John W. Hyatt (1837–1920). Formulated Celluloid, a camphor plasticized nitrocellulose, usually accepted as the first syn-

President Herbert Hoover, Henry Ford, and Harvey Firestone at Fort Myers, Florida at the time of Edison's 82nd birthday, February 11, 1929

thetic plastic; developed injection molding and the first multicavity mold; invented the Hyatt roller bearing.

Charles B. Dudley (1842–1909). Chief chemist for years for the Pennsylvania Railroad Company; helped form and was first president of the American Society for Testing Materials; made an exhaustive study of all materials going into railroad construction.

Harvey W. Wiley (1844–1930). Chief chemist of U.S. Department of Agriculture for 30 years; responsible for first pure food law; fostered the sugar beet industry; improved and standardized methods of agricultural chemical analyses.

Charles F. Brush (1849–1929). Worked first as a chemist and then primarily in electric field; developed arc lights which could be turned on by remote control using electromagnets; pioneered arc lighting for city streets; Brush, as well as Faure, improved the lead storage battery of Planté by applying active lead compounds to the plates.

Edward Weston (1850–1936). Trained in medicine but devoted himself to chemistry and physics; active in electric lighting and the nickel-plating industry; manufactured dynamos designed for electroplating; Weston Electric Instrument Company became world famous; developed cadmium amalgam standard cell which has been standard for electromotive force since 1901; obtained 309 patents.

Herman Frasch (1851–1914). Contributed much in the early refining of petroleum, particularly method of desulfurization by means of metal oxide which could be recovered, burned, and reused; developed and put into practice method of mining sulfur in porous deposits by melting it with superheated water and bringing it to the surface by compressed air.

William H. Nichols (1852–1930). Formed his own chemical company at age of 18; contributed to technologies of making sulfuric acid and the electrolytic refining of copper; by mergers formed General Chemical Company in 1899 and Allied Chemical and Dye Corporation in 1920; said to be the "first captain of industry in America to capitalize the value of research."

George Eastman (1854–1932). Pioneered the replacement of wet glass photographic plates with those coated with dry gelatin–silver halide emulsion; paper and later plastic film replaced the glass; developed low cost cameras for using roll films; his simplification of the photographic art led to a tremendously successful business.

George W. Carver (1861–1943). Although primarily an educator (Tuskegee Institute), Carver was also a leading chemurgist; pioneered the growing of peanuts and sweet potatoes in the South; developed outlets for peanut hulls and pine cones for making wallboard and other building products; showed farmers how to make paints from clays.

Charles L. Reese (1862–1940). Chief chemist of New Jersey Zinc Company; in 1902 director of Du Pont's Eastern Laboratory; pioneered work on blasting caps and contact process for making sulfuric acid; director of Du Pont's chemical division beginning in 1911.

Charles M. Hall (1863–1914). Eight months after graduation from Oberlin College invented process for manufacturing aluminum by electrolysis of alumina dissolved in fused cryolite (Na_3AlF_6); became vice-president of Pittsburgh Reduction Company (now Aluminum Company of America).

Herbert H. Dow (1866–1930). Founder of Dow Chemical Company; first to manufacture bromine and bromine salts in the U.S.; pioneered electrolytic production of chlorine and caustic soda, bleaching powder, chloroform, and certain agricultural chemicals.

Harvey S. Firestone (1868–1938). Organized rubber manufacturing business in 1896; leader in attempt to grow *Hevea* closer to United States than the Far East; introduced balloon tire in 1923; foremost in the industry in establishing research laboratories.

Willis R. Whitney (1868–1958). Representative of the new breed of highly trained chemists entering industry in the early 20th century; first director of research at General Electric; contributed to improvements of carbon filaments for incandescent lighting.

Of these contemporaries, Bell and Westinghouse were elected to the "Hall of Fame." Admission to this highly exclusive group (only 93 to date) is indeed a certificate of excellence (3). Bell and Westinghouse were elected, as was Edison, under the classification of inventor. All inventors, of course, are not chemists.

Two additional outstanding American scientists, who made their marks in industry prior to World War I, were Elihu Thomson (1853–1937) and Charles Steinmetz (1865–1923). Thomson joined Edwin Houston to form the Thomson-Houston Electric Company in 1880. It became a part of General Electric in the merger of 1892. Steinmetz, at General Electric, worked primarily on theoretical aspects of alternating current. Although both Thomson and Steinmetz made many inventions and pioneered new things, their achievements were strictly electrical in nature and are not pertinent to this discussion.

In the previous chapters Edison's technical projects have been emphasized where chemistry played a significant part. Projects, significant but involving little chemistry, include many inventions in telegraphy, dynamos, the electric railway, the motion picture, and the discovery of thermionic emission (Edison Effect). Chemical projects, of a smaller magnitude than those discussed in previous chapters, include the preparation and testing of hundreds of metallic compounds as fluorescents on x-ray screens, contributions in the invention and manufacture of the fluoroscope, the carbon transmitter for the telephone, microphone, etc., various improvements in electrochemistry, and various projects in refining metals from their ores.

By way of recapitulation, Edison's principal contributions in the field of chemical research and development are:

(a) The development of carbon fibers.

(b) First demonstration that carbon could be heated to a light yellow color (over 2,000 °C) without disintegration or severe volatilization even *in vacuo*.

(c) The first manufacture of a practical device containing a very high vacuum enclosure capable of maintaining that vacuum for months or even years.

(d) Major advances in the fields of electrical insulation for indoor, outdoor, and underground installations.

(e) Development of the first commercial alkaline storage battery and pioneered the manufacture of the nickel–iron battery.

(f) Several advances in the field of electrochemistry, including method for making nickel flakes of minute thickness and a method for making zinc and other metal sheeting in a continuous manner.

(g) First observation of the strong adsorption of gases by platinum metal.

(h) Development of several new techniques for concentrating magnetite ores by magnetic separation.

(i) First preparation of iron powder commercially by direct hydrogenation; he manufactured and sold high purity iron to chemical reagent suppliers.

(j) Pioneering work in the processing and molding of plastics.

(k) Pioneering the development and use of rubber-fabric conveyor belting.

(l) First use of foam concrete for structural purposes.

(m) Several firsts in the manufacture of organic chemicals during World War I.

(n) First determination that goldenrod leaves contain appreciable concentrations of rubber; he determined rubber contents of literally thousands of plants.

It is not the purpose of the author to attempt to weigh the evidence pro and con as to whether Edison contributed more or less than many recognized outstanding American chemists in the era prior to World War I. Adequate evidence has been presented that he was a chemist and made substantial contributions in the field. However, the fact remains there is substantially nothing in chemical textbooks or other chemical treatises about his work.

The Perkin Medal is awarded annually by the American Section of the Society of Chemical Industry "for the most valuable work in applied chemistry." The award may be made to any chemist residing in the United States for work done at any time during his career. By the time of Edison's death in 1931, nine of Edison's contemporaries had received this award: Acheson, Hall, Frasch, Hyatt, Weston, Baekeland, Whitney, Dow, and Little. Why wasn't Edison so recognized? Certainly his work had a more "chemical flavor" than that of Weston. The answer may be that he was offered the award and refused it. One of the unofficial obligations in granting the award is that the recipient address the gathering at the time the award is given, describing his contributions on which the award is based. Edison rarely made a speech, particularly after his hearing loss became severe.

I do not wish to give the impression that Edison did not get numerous awards and died a bitter man because of it. He received 33 medals of recognition including the French Legion of Honor, the Franklin Medal, and the U.S. Congressional Medal. He was given honorary memberships in 16 scientific and engineering organizations. He received seven honorary degrees including a Ph.D. from Rutgers University, an LL.D. from New York State University, and a Doctor of Science degree from Princeton. However, the nearest any of these could be called a recognition in chemistry was an honorary membership in the American Electrochemical Society, granted in 1928. Despite all the doctoral degrees which he received, he never allowed anyone to address him as Dr. Edison.

There are many stories about Edison's refusing to appear in person to accept various honors and awards. His LL.D. degree from New York State University was conferred over the telephone by Dr. John H. Finley, Chancellor, speaking from Albany to Edison at his laboratory in West Orange. When he was invited to Washington to receive the Special Congressional Medal of Honor, voted to him by the U.S. Congress in 1928, he sent word that he was too busy to make the trip. However, Senator Edge (N.J.), who had sponsored the bill for the award, did not give up easily. He prevailed upon Secretary of the Treasury Andrew Mellon to come to the Edison Laboratory and make the presentation there.

It is more than likely that the reason for such obstinacy was Edison's hearing problem. Dr. Charles L. Edgar, president of the Edison Electric Illuminating Company of Boston at the time, has related the following incident which happened because of Edison's deafness (4):

This infirmity sometimes had its ludicrous side. I remember about twenty or more years ago, when a dinner was given to

Lord Kelvin, Sir William Thomson, at the Waldorf in New York, Mr. and Mrs. Edison were present, and in the speech-making which took place some very complimentary remarks were made by the chairman regarding Mr. Edison. Everybody, including Mr. Edison, applauded vociferously. His wife leaned over, pulled down his head, and whispered something to him. He immediately blushed, sank down in his chair and almost disappeared under the table.

Just because Edison received no awards in chemistry is no explanation why his work should have been ignored by the chemical press. The reason for this is probably because nearly all books dealing with chemistry, particularly before World War I, have been written by academic people. Their source of new knowledge in the field is the chemical literature. Edison's only publications in the technical literature appear to have been his article published by AAAS in 1879 and cited in Chapter 2 and the joint article he had with Birkinbine in a mining journal as cited in Chapter 5. This does not mean that Edison did not publish. His primary publications were in the form of patents, and they were extensive. Edison's patents are clear, concise, and are almost invariably accompanied by well defined dia-grammatic sketches. In comparison with present-day patents, those of Edison are characterized by their brevity. It is unfor-tunate that even today most teachers and academic researchers ignore the patent literature. Theodore Edison also feels that many of his father's scientific achievements lie "buried" in his patents. The former's statement relative to this follows (5):

Even a brief discussion of Mr. Edison's work would not be complete without some mention of his contact with patents. He obtained a record number himself, and in the course of getting them I feel sure that he must have picked up a great deal of useful information. I think it is too bad that the formal legal language used in patents is so hard to understand, because patents contain a wealth of information that may never reach the ordinary technical publications. A lot of patents describe results obtained by so-called practical men who may have had very little formal education, but who nevertheless may have developed valuable ideas.

There is one first which Edison achieved in the scientific field on which there can be no disagreement—that is, the establishment of the first industrial research laboratory in the United States. In fact, both Charles Kettering, the noted inventor, and General David Sarnoff, renowned executive of the RCA Corporation, have called this Edison's greatest invention. In the case of the laboratory at Menlo Park, one might disagree that it was the prototype of our present concept of an industrial research laboratory since Edison had no industry at that time and much of the work was being supported by outside groups. However, in the case of the Edison Laboratories at West Orange, there is no such question. Research projects there were supported by the Edison industries in the neighboring area as well as by royalties from patents licensed, work was being done on improving the current factory operations as well as developing new products and new processes, and developments from the test tube to plant-scale operation were being pioneered.

Some may contend that Edison's research was different from present-day practice in that his was a one-man type of operation. This was true to a degree at Menlo Park. However, even there Dr. Alfred Haid was in charge of preparing chemicals which could not be bought, purifying some of the purchased chemicals, and working on the analysis and recovery of metals from ores. Francis Upton, physicist and mathematician, was responsible for the design of much of the electrical equipment besides the light. At West Orange Edison had what we would call section heads in charge of the various research groups. The laboratory was organized in such a way that if Edison were absent for months, as in the case of some of his winter "vacations" in Fort Myers, the work would continue.

Noted Edison employees included Jonas Aylsworth, chemist, Dr. Arthur E. Kennelly, who succeeded Upton as Edison's theo-

Reproduced from F. T. Miller's "Thomas A. Edison"

Edison with a group of motion picture executives. This picture illustrates that Edison could enjoy such a gathering and apparently be the "life of the party."

retician and later became famous as a professor of physics at Harvard, John Howell, lighting engineer, T. C. Martin, electrical engineer, and hosts of others who moved into responsible positions in the various Edison industries or with other industrial companies (6).

It is surprising that many present-day chemists do not know of Edison's pioneering of the industrial research laboratory. Just a few years ago I heard a Perkin Medalist state in his award address that the first industrial research laboratory was organized by General Electric and the first relating to industrial chemistry was that of Kenneth Mees at Eastman Kodak. Mees did not begin work with Eastman until 1913, and for many years his research was wholly on photography. General Electric set up a research laboratory entirely separate from factory and sales in 1900. Dr. Whitney of M.I.T. was hired to organize that laboratory, and he began work at Schenectady on a part-time basis with one assistant. In 1904 Whitney moved from Cambridge to Schenectady, and three additional chemists were brought in from M.I.T. Work centered on improving the incandescent lamp, and G.E. hired chemists for this project—not electrical engineers or physicists.

Du Pont became a corporation in 1899. In 1902 their Eastern Laboratory was established to work on improved explosives. The following year their Experimental Station was organized and housed in an old cotton mill on the Brandywine. Initial research at the Station was directed to products closely related in chemical composition to the nitrocellulose explosive. These products included artificial leather, plastics, photographic film, and nitrocellulose lacquers.

It is unfortunate that Edison did not have more time for writing and that his deafness caused him to shun the speaker's platform. When speaking with his workmen, he usually used the language of the uneducated so they wouldn't think he was putting on any airs. With his college-trained employees, he delighted in appearing as a country bumpkin, with liberal use of such terms as ain't, git, somep'n, and the like, in an attempt to impress upon them that he got where he was by hard work alone—hoping, of course, that they would take the hint. However, when he chose, he could write and speak with imaginative elegance. Following is a paragraph from his diary kept over a period of 10 days in July 1885, which covered the time while on a trip with his daughter Dot to visit the Ezra Gillilands at their summer home in the Woodside Park section of Winthrop, Massachusetts. The quote is from his July 13 entry dealing with the stopover in New York City (7):

Went into Scribner and Sons on way up, saw about a thousand books I wanted. Right off Mind No. 1 said, Why not buy a box full and send to Boston now. Mind No. 2 (acquired and worldly mind) gave a most withering mental glance at Mind No. 1 and said, You fool, buy only two books. These you can carry without trouble and will last until you get to Boston. Buying books in New York to send to Boston is like "carrying coals to Newcastle." Of course I took the advice of this earthly adviser. Bought Aldrich's story of a bad boy, which is a spongecake kind of literature, very witty and charming, and a work on Goethe and Schiller by Boynsen, which is soggy literature. A little wit and anecdote in this style of literature would have the same effect as baking soda on bread—give pleasing results.

One of Edison's last speeches, characteristically short, was at the 50th anniversary of the invention of the first practical incandescent lamp. The elaborate affair was held at Dearborn, Michigan in conjunction with the dedication of the restored Menlo Park laboratory in Greenfield Village and the Edison Institute, both having been recently completed by Henry Ford. Light's Golden Jubilee was celebrated in many cities throughout the world. The U.S. Government issued a memorial postage stamp in honor of Edison. President Hoover, Madame Curie, and

Picture taken in Edison's study at Glenmont a few months before his death,
October 18, 1931

about 500 other notables were present at the official dinner held in the Independence Hall replica on the evening of October 21, 1929. Owen D. Young, president of General Electric, served as master of ceremonies. A congratulatory message from Professor Albert Einstein, then in Germany, was received *via* radio. When Edison arose to deliver his appreciation, a great silence prevailed. In a voice filled with emotion, he spoke as follows:

I am told that tonight my voice will reach out to the four corners of the world. It is an unusual opportunity for me to express my deep appreciation and thanks to you all for the countless evidences of your goodwill. I thank you from the bottom of my heart.

I would be embarrassed at the honors that are being heaped upon me on this unforgettable night, were it not for the fact that in honoring me, you are also honoring that vast army of thinkers and workers of the past, and those who will carry on, without whom my work would have gone for nothing.

If I have spurred men to greater efforts, and if our work has widened the horizon of man's understanding even a little and given a measure of happiness in the world, I am content.

This experience makes me realize as never before that Americans are sentimental, and this crowning event of Light's Golden Jubilee fills me with gratitude. I thank our President and you all.

As to Henry Ford, words are inadequate to express my feelings. I can only say to you that, in the fullest and richest meaning of the term, he is my friend.

Good-night.

References and Appendices

The following references and supplementary information are itemized to correspond to the individual citations in each chapter. In the interest of brevity, the name of the patentee has been omitted for the cited patents of Thomas A. Edison.

Chapter 1 (pp. 1-28)

(1) Mitts, Dorothy, "That Noble Country," Dorrance & Company, Philadelphia, 1968. Miss Mitts, a former newspaper writer of historical topics, deals with the history of the St. Clair River region in this interesting book. Chapter XXX, pp. 199-216, is entitled "The St. Clair District's Most Celebrated Pioneer: Thomas Alva Edison."

(2) At the end of the 18th century in Europe there were several semaphore systems for transmitting messages over considerable distances. A system in France from Paris to Toulon covered 475 miles and had 120 manned stations.

(3) Nitrated cotton and nitroglycerin had been synthesized earlier by Europeans and were known to be explosives. Although Nobel had found by 1866 how to handle nitroglycerin safely by adsorption on kieselguhr, and cellulose trinitrate by solution in amyl acetate to form a gel, the du Ponts did not manufacture either of these more effective explosives until the 1880's. Much later they diversified into chemicals by acquisition and research.

(4) The lack of interest in organic chemistry in the colleges and universities of America at this time existed despite the fact that Wöhler had carried out his famous synthesis of urea, an organic, from a wholly inorganic material, ammonium cyanate, in 1828. Perkin in 1856 had synthesized mauve, a purple dye, by the oxidation of toluidine. Perkin's so-called aniline dye was introduced to the New York area in 1860 at a price of $327 per lb. Although this and other synthetic dyes to follow, primarily from Germany, found a ready market in America, none was manufactured here up to World War I.

In 1865 Kekulé's work with hydrocarbons resulted in his postulation of the closed ring structure for benzene and other aromatic hydrocarbons. This discovery failed to spark work in American colleges on hydrocarbons even

345

though we were the leading petroleum producer of the world and soon a leading coke manufacturer. America had no coal tar chemical industry up to World War I, at which time Thomas Edison played a major role in making such essential organic chemicals available in the United States (Chapter 8).

(5) Full citation: U.S. Department of the Interior, National Park Service. Edison National Historic Site; hereafter abbreviated as shown.

(6) Michael Faraday (1791–1867) was one of the great experimental scientists of all time. He discovered and explained electromagnetism and is considered the father of the dynamo. He had worked as an assistant to Sir Humphrey Davy who, in turn, was the first to bring chemistry and electricity together and to show that there were positive and negative ions when an electrolyte is dissolved in water. Davy prepared sodium and potassium by electrolysis of their molten salts and discovered chlorine and iodine.

Faraday developed electrochemistry to an exact science by showing that in electrolysis the amount of decomposition depends on the quantity of electricity—not its intensity—and the chemical equivalent weights of the constituents. Faraday's first work and much of his experimental studies dealt with chemistry. He was the first to isolate benzene, obtaining it from the decomposition products of whale oil in 1825 and naming it bicarburet of hydrogen. It was not until 1845 that it was isolated from coal tar by A. W. Hofmann.

The titles of Faraday's "Principal Works" illustrate his activity in the chemical field:
  (a) Chemical Manipulation (1827)
  (b) Experimental Researches in Electricity (1844)
  (c) Experimental Researches in Chemistry and Physics (1859)
  (d) Lectures on The Chemical History of a Candle (1861)
  (e) On the Various Forces in Nature (undated)
Actually (d) is not as limited as it sounds; for example, one lecture dealt with the platinum family of metals.

(7) Matthew Josephson, author of "The Robber Barons," was no doubt in a position to have a great deal of background information on the struggle of the financiers for control of the telegraph business. He covers this phase of American history and how it affected Edison in Chapters VI and VII of his book "Edison," McGraw-Hill Book Company, New York, 1959.

Chapter 2 (pp. 29-70)

(1) George S. Ohm, a German physicist, in 1827 published a pamphlet on electricity which included the formulation known today as Ohm's law. Ohm deduced that the current in a wire is proportional to the potential difference at the ends of the wire. For a given circuit, this ratio is constant and is known as the resistance. In modern terminology, $R = E/I$, and using today's electrical units, 1 ohm = 1 volt per ampere. Expressing Ohm's law as $E = IR$, it is readily apparent that a high resistance lowers the flow of current for a given voltage.

Electrical energy is work and is a product of force times distance; moving 100 lbs upward 10 ft takes 1,000 ft-lbs of energy. The equivalent in electricity is volt-amperes, and when the product is 1, the amount of work done is 1 watt.

(2) As everyone knows who has handled a string of electric lamps on a two-wire electric cord, there are two types of connections to the lamp sockets. In one type all lights go out if a bulb fails; in the other, the remaining bulbs stay lit after one or more lights burn out. In the former type of circuit, series wiring, the current has one continuous path, that for the last lamp must pass through all the other lamps. With parallel wiring, one can consider the two wires as being the sides of a ladder, never meeting except *via* the crossbars. Each lamp is actually on such a crosswire, and if a lamp fails, it does not affect the others. The amount of current flowing across each "rung" of the circuit is so small that the lamp near the end of the "ladder" has almost the same differential of electromotive force as does the first lamp.

(3) When a gray or black object is heated it goes through the following color changes: dull red, bright red, orange, yellow, white, and finally blue. The final two colors are based on theory only and are never achieved in practice. The higher one can attain in the thermal color scale with incandescent lighting, the greater is the efficiency of converting electrical energy to light.

(4) Greenfield Village at Dearborn, Michigan was built by Henry Ford to preserve many features of American life of the late 19th century. An important part of this restoration is Edison's Menlo Park laboratory and auxiliary buildings, as well as the Jordan Boarding House across the street and Edison's small laboratory from Fort Myers. Adjacent to the Village is the vast Henry Ford Museum, housed in an enlarged replica of Independence Hall of Philadelphia. Literally thousands of exhibits portray the accomplishments of American industry and invention. The corporate title of Greenfield Village and the Henry Ford Museum is The Edison Institute.
Transplanted Menlo Park was indeed done in an authentic manner. Robert Koolakian, Associate Curator, maintains the authenticity by such things as "manufacturing" on the site carbon-filament light bulbs used for lighting. At the time of the dedication of the Village and Museum in October 1929, Thomas Edison observed that the restored laboratory was 99% perfect—the one percent error being that it was too clean. A visit to Dearborn to view the various restorations and exhibits is an exciting experience.

(5) Jehl, Francis, "Menlo Park Reminiscences," published by The Edison Institute, Dearborn, Michigan, Vol. I, 1934; Vol. II, 1938; Vol. III, 1941.

(6) Mendeleyev used his periodic arrangement of elements to predict new chemical facts. He predicted the existence of the elements gallium, scandium, and germanium before their discovery. He questioned the correctness of certain "accepted atomic weights" on the ground that they did not correspond with his periodic law, and here also he was vindicated by subsequent investigations.

(7) Edison, T. A., *Sci. Amer.* (1879) **41**, 216.

(8) Upton, Francis, *Proc. Amer. Assoc. Advan. Sci.* (1879) 178-84.

(9) Edison, T. A., *Proc. Amer. Assoc. Advan. Sci.* (1879) 173-77.

(10) Howell, J. W., Schroeder, Henry, "History of the Incandescent Lamp," Maqua Company, Schenectady, New York, 1927.

(11) Jehl, Francis, "Menlo Park Reminiscences, II," p. 811, Edison Institute, Dearborn, Michigan, 1938.

(12) *The Operator* (July 15, 1880) **3**, 4.

347

Chapter 3 (pp. 71-97)

(1) Williams, David, "The Diamond Jubilee of Consolidated Edison Underground Cable," p. 3, a monograph published by the Consolidated Edison Company, New York, New York. This treatise was based on a paper presented before the Wire and Cable Section of the National Electrical Manufacturers Association at Atlantic City, New Jersey on November 14, 1956.

(2) The so-called feeder-main system was covered by Edison's patent U.S. 264,642, applied for August 4, 1880 and issued September 19, 1882. To quote from its specification:

> This invention relates to a method of equalizing the tension or "pressure" of the current through an entire system of electric lighting or other translation of electric force, preventing what is ordinarily known as a "drop" in those portions of the system the more remote from the central station . . .

Wires transmitted current from the dynamos to central points of short sections of main lines from which conduits to individual customers were fed. The main lines were of relatively high cross-section, as well as short in length, to avoid voltage drop from the midpoint to either end. All lines to customers were 110 volts, which was the highest that Edison felt could be safely handled in the home. The 110-volt standard has continued to present day.

The feeder system made the control of individual lights and motors feasible without the use of excessive quantities of copper. When this system was demonstrated at a meeting in London, the noted physicist, Lord Kelvin, was asked why no one had ever thought of such a simple system before. He replied, "The only answer I can think of is that no one else is Edison."

In connection with the design of the circuits for the first central electric station located at 255-57 Pearl Street in lower Manhattan, a miniature model of the approximate one square mile to be serviced was laid out on a large table at the Menlo Park laboratory. From an earlier survey it was known approximately how many lamps and how much power equipment would be used on each street. Small resistance coils simulated the various power requirements, and it was determined experimentally, using batteries for the source of current, where the best locations would be for the power station and the feeder lines.

(3) Edison realized that if he could use a voltage above 110 in the feeder and main lines, he could decrease the diameter of these wires. Since in the circuits running to customers he was already substituting voltage for amperage to the maximum level safety allowed, it meant reducing voltage to the wiring leading to the lamps and other electrical devices. This was before the transformer, which today is used to decrease voltage at the street before electrical power is delivered to the customer. However, by means of the so-called three-wire system, U.S. Patent 274,290 (March 20, 1883), Edison was able to cut his line voltage to half at the site of the user.

It is apparent from the diagram below that the three-wire system comprises two circuits combined into one system. When all eight identical lamps (L) are on, the current from generators G^1 and G^2 passes through the positive conduit (P) and returns by the negative (N). Each generator produces 110 volts making 220 volts in the main line, but since the current passes through two lights, there are only 110 volts at the lights. The third wire, PN, under these conditions is neutral as far as the generators are concerned. However, if one lamp is extinguished on the positive side of the circuit (P), the system is thrown out of balance. There is an excess of power on the positive side equivalent to that previously drawn by the extinguished lamp. This excess immediately flows into the neutral wire and serves to furnish the power for the fourth light on the lower level. If the extinguished lamp had been on the negative side of the circuit (N), the neutral wire would for the moment carry excess current back to the source of energy.

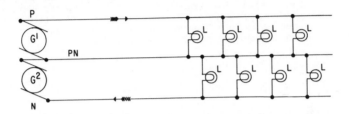

At nearly the same time Edison was developing the three-wire system, it was conceived independently by both Dr. John Hopkinson of England and Dr. Werner Siemens of Germany. Hopkinson was doing some work for the English Edison Company at the time and no doubt had the benefit of considerable know-how from the parent company.

(4) Williams, David, *op. cit.* (Ref.*1*), p. 3:
   The earliest references to the commercial use of copper as a conductor of electricity indicate that in 1811, two Germans— Soemering and Schilling—first used copper wire in a rubber insulated submarine cable laid in the bed of the Isar River near Munich, Germany. In the years 1812–1815, Paul Canstadt, Baron von Schilling used a copper wire insulated with a solution of india-rubber and varnished, for detonating mines under the Neva and Seine Rivers. In 1838 some development work on rubber insulated electical wires was done in England.
   The Wire and Cable industry of the United States may be said to have been born with the invention of the telegraph by Samuel F. B. Morse in 1837. Bare copper wire was used on the first telegraph lines, but was replaced with iron wire of less conductivity, but of greater tensile strength. However, it soon became evident that reliability of service required that the wires be covered or insulated. The early forms of insulation consisted of cotton cloth saturated with beeswax, paraffin, or gum shellac and various compounds of gutta percha, and unvulcanized rubber. Morse used an unvulcanized rubber-insulated submarine cable between the Battery and Governor's Island, N. Y. in 1841.
   The first commercial production of rubber insulated wire in the United States was started in 1866 by The Kerite Company of Seymour, Conn., which had originally been founded by Austin Goodyear Day in 1854 for the production of hard rubber products. A dry paper-wrapped wire (the forerunner of the modern telephone cable) was made in 1872 by John H. Wortendyke of Richmond, Va. for use on the electric bell system in his home,

but it was not commercialized until E. O. McCracken, who had obtained a patent on September 2, 1884, formed the Norwich Insulated Wire Company in 1886. It was subsequently (1891) acquired by the National Conduit & Cable Company (a predecessor of Anaconda Wire & Cable Company).

The first recorded installation of underground wire for electric arc lighting in the United States took place on the Campus of Cornell University in 1875, when electrical engineering students used copper wire lapped with two layers of muslin secured with a spiral cord and impregnated with beef tallow obtained from a local butcher. The insulated wire was drawn into an iron pipe which was then buried in the ground, and connected the arc lamps to the first dynamo built in the United States—that of their Professors W. A. Anthony and G. S. Moler.

(5) Jehl, Francis, "Menlo Park Reminiscences, II," pp. 723-24, Edison Institute, Dearborn, Michigan, 1938.

(6) Silverberg, Robert, "Light for the World," Van Nostrand, Princeton, New Jersey, 1967.

(7) Josephson, Matthew, "Edison," pp. 364-366, McGraw-Hill, New York, 1959.

(8) Tate, A. O., "Edison's Open Door," pp. 260-266, E. P. Dutton, New York, 1938.

(9) Speiden, Norman, Edison National Historic Site, West Orange, New Jersey, private communication, 1969.

(10) Josephson, M., *op. cit.,* pp. 271-272.

Chapter 4 (pp. 98-137)

(1) Dyer, F. L., Martin, T. C., "Edison, His Life and Inventions, I," pp. 206-208, Harper, New York, 1910.

(2) Read, Oliver, Welch, W. L., "From Tin Foil to Stereo," p. 6, Howard W. Sams, Indianapolis, 1959. The reader interested in the history of sound recording and reproduction will find this book an excellent reference. The 524-page treatise includes numerous sketches and photographs. The author wishes to acknowledge the use of this source of information. Another good reference is: Gelatt, Roland, "The Fabulous Phonograph," J. B. Lippincott, New York, 1955.

(3) The following is based on private communications from N. R. Speiden and H. S. Anderson, March 1970. Several books, including that of Read and Welch (Ref. 2), have carried a photograph of a rough laboratory sketch revealing the basic recording–playback technique on which is written "Kruesi make this, Edison," dated "Aug 12/77." Francis Jehl also published this sketch in his first volume of "Menlo Park Reminiscences." However, in a visit to the West Orange Laboratory in 1938, Jehl pointed out to Speiden that this sketch had been published twice before but with no writing on it. It had so appeared in the book "The Life and Inventions of Thomas A. Edison" by the Dicksons in 1894 with the acknowledged source as J. U. Mackenzie, Edison's old telegrapher friend. On checking

with Mackenzie's daughter, Mrs. Naomi Mackenzie Chaffee of Montclair, New Jersey, Speiden learned that the sketch was casually made by Edison while talking to her father, and Mackenzie had retained it. The sketch did not leave Mackenzie's possession until loaned to the Dicksons in 1894. It subsequently fell into the hands of the Edison Phonograph people. In 1917, at the time the 40th anniversary of the phonograph was celebrated, this sketch was "doctored up" to indicate its having been given to Kruesi on August 12, 1877. The original sketch given to Kruesi could not be found, and based on a study of the notebooks the date of August 12 appeared authentic.

About 20 years after Jehl's detection of the "fraud," Charles Batchelor's daughter, Emma Batchelor, gave the Edison National Historic Site many records which had been kept by her father at Menlo Park. In a "Day Book" for 1877 were the following entries:

December 4th—Kruesi made the Phonograph today.
December 6th—Kruesi finished the Phonograph.
December 7th—We took the Phonograph to New York to the
office of Scientific American.

Subsequent search showed a sketch in one of Edison's notebooks showing the design of the December 6 machine, dated November 29, 1877.

(4) Edison, T. A., "The Phonograph and Its Future," *North American Review* (June 1878).

(5) From a handwritten copy in Edison's laboratory notebook of November 15, 1887.

(6) Meadowcroft, W. H., "The Boy's Life of Edison," pp. 308–10, Harper, New York, 1911.

(7) U.S. Patent 400,648 (April 2, 1889).
U.S. Patent 406,569 (July 9, 1889).

(8) U.S. Patent 406,576 (July 9, 1889).

(9) U.S. Patent 414,761 (November 12, 1889).

(10) U.S. Patent 382,419 (May 8, 1888).

(11) U.S. Patent 384,584 (October 18, 1892).

(12) Phonograph Formula Book, courtesy of the Edison National Historic Site, West Orange, New Jersey.

(13) Welch, W. L., Syracuse University, Syracuse, New York, private communication, September 1970.

(14) Jewett, F. B., "Edison's Contributions to Science and Industry," *Science* (January 15, 1932) **75**, 67.

(15) Millikan, R. A., "Edison as a Scientist," *Science* (January 15, 1932) **75**, 69.

Chapter 5 (pp. 138-177)

(1) A paper entitled "The Concentration of Iron Ore," presented before the American Institute of Mining Engineers by John Birkinbine and Thomas Edison in New York City in February 1889. It was subsequently published

in the *Transactions of the American Institute of Mining Engineers* (1888–89) **17**, 728. Birkinbine was a mining engineer from Philadelphia, Pennsylvania who apparently had been retained by Edison as a consultant.

(2) U.S. Patent 228,329 (June 1, 1880). This was the first of 16 patents obtained by Edison on various arrangements and equipment for magnetic separation of magnetite and other magnetic ores, such as pyrites. No. 676,840, issued June 18, 1901 involves the use of an endless belt to carry the ore over a magnetic pulley—a technique widely used today.

(3) Birkinbine, J., *Trans. AIME* (1890–91) **19**, 656.

(4) Ball, C. M., *Trans. AIME* (1890–91) **19**, 187.

(5) Smock, J. C., *Trans. AIME* (1891) **20**, 224.

(6) *Ibid.*, p. 585.

(7) "The Edison Concentrating Works," *Iron Age* (October 28, 1897) p. 1.

(8) U.S. Patents 672,616 and 672,617 (April 23, 1901). Edison once said that his patents covering rock crushers were his only patents which were not violated by others. He no doubt meant those patents which became industrially important. Edison, like all prolific inventors, took out patents on processes and products which failed to be carried to a commercial scale.

(9) U.S. Patent 648,933 (May 8, 1900).

(10) U.S. Patent 675,057 (May 28, 1901).

(11) U.S. Patent 648,934 (May 8, 1900).

(12) Dyer, F. L., Martin, T. C., "Edison, His Life and Inventions," Vol. 2, pp. 491-92, Harper, New York, 1910.

(13) U.S. Patents 465,251 (December 15, 1891) and 485,840 (November 8, 1892).

(14) U.S. Patent 661,238 (November 6, 1900). The novelty of this so-called bricking machine was the use of a rotatable mold carrier containing three cavities. The plungers worked simultaneously, and progressively increased pressures were applied in the three compression stages. This is one of Edison's longest patents, consisting of seven pages and having 20 claims.

(15) *Hewitt-Robins News of Industry* (1961) **7** (1), 6.

(16) Spilsburg, E. G., *Trans. AIME* (1897) **27**, 452.

(17) Birkinbine, J., *Trans. AIME* (1897) **27**, 519.

(18) Tate, A. O., "Edison's Open Door," p. 280, E. P. Dutton, New York, 1938.

(19) McInerney, Robert, Alan Wood Steel Company, Conshohocken, Pennsylvania, private communication, 1969.

(20) Alan Wood used both dry and wet methods in the magnetic separation. Aqueous methods were favored to cut down on the dust problem and its inherent health hazards.

The major part of the high grade ores of Mesabi Range has been depleted. A large formation of 25–30% iron in the form of magnetite called taconite is now being concentrated and agglomerated. The overall upgrading of such ores is now called beneficiation. The process involves crushing the open mined ore to a maximum size of ¾-inch and then grinding in rod or ball mills. This is done in large rotating drums containing the ore and the balls or parallel steel bars, as well as enough water to provide a 25–60% slurry of the ground particles. Concentration is usually carried out in a three-stage system in which the slurry of ore as small as 200 mesh in particle size is contacted with moving drums passing over fixed magnets. With the recent development of improved magnets in the form of so-called ferrites, such as that of barium (which are actually ferrates since the iron has a valence of 3), these are usually used instead of electromagnets. However, Alan Wood Steel Company used electromagnets in their concentration plants in New Jersey.

The concentrated taconite ore contains 60–66% iron and 2–8% silica. The slurry from the magnetic concentrator is filtered, and the filter cake goes to a balling drum with spiral grooves. Here it is mixed with a small amount of finely divided coal and bentonite clay and rolled into balls $\frac{3}{8}$ to $\frac{3}{4}$ inch in diameter. These are dried and then baked at 2,200°–2,500°F. In an alternate process the fine concentrate is mixed with binder and formed into pellets by passing between two rotating rolls which contain pockets into which the ore is forced under pressures of 10–30 tons per square inch. The pockets have a curved base, and usually a pair of pockets mesh together resulting in a "double pellet" with all curved surfaces. Spherical and elliptical shaped agglomerates are preferred to those having flat surfaces since they pack more loosely in the blast furnace. From the standpoint of shape, Edison's briquettes had been good in that they were circular, but poor in that they had two flat sides.

Although pelletizing and briquetting are more expensive than sintering for agglomerating finely divided ores, they are being increasingly used since the iron trade prefers such regularly shaped ore agglomerates. Thus, Edison's decision to go the briquette route for agglomerating his concentrate was a farsighted one.

Although magnetite makes up only a minor proportion of the total world production of iron ores, it may be economical to reduce hematite to Fe_3O_4 so that it can be purified by magnetic separation. For example, Montecatini, the large manufacturer of chemicals, plastics, and other synthetics in Italy, carries out the following steps in purifying a hematite ore. The fluid bed technique is used in Steps 1 and 2:

(1) The ore is roasted (oxidized) to convert any iron sulfide to Fe_2O_3 and the sulfur recovered as sulfuric acid.

(2) The Fe_2O_3 is reduced to Fe_3O_4 and cooled.

(3) The Fe_3O_4 is concentrated by magnetic separation.

Since the beneficiation of iron ore is becoming such an increasingly used practice, the availability of such high purity concentrates has led to direct reduction processes, thus bypassing the blast furnace operation. The fluid bed technique is particularly applicable for reducing iron oxides to metallic iron using hydrogen and/or carbon monoxide. By using the reducing gas under pressure, temperatures as low as 600°F can be used with resulting savings in fuel. The finely divided metal obtained is finding increasing uses as such, but is usually pelletized and marketed for making steel in competition with pig iron and scrap metal.

Thus, the approach of concentrating, removing harmful impurities, and forming ores into agglomerates at the mines, rather than operating the blast furnaces in such a way so as to utilize low purity ores, is now the accepted practice in the iron and steel industry. Certainly Edison was one of the early pioneers in this development.

(21) Dyer, F. L., Martin, T. C., "Edison, His Life and Inventions," Vol. 2, pp. 502–04, Harper, New York, 1910.

(22) "Edison: Town that Disappeared" was the title of two articles which appeared in the Sunday editions of the *New Jersey Herald*, Newton, New Jersey, January 21 and 28, 1968.

(23) "Ogdensburg Golden Anniversary," An 11-page booklet published by the Borough of Ogdensburg, New Jersey in 1964. It can be consulted at the Sussex County Library, Branchville, New Jersey.

(24) Milstead, C. R., U.S. Steel Corporation, New York, New York, private communication, December 1970.

Chapter 6 (pp. 178-202)

(1) The Delaware, Lackawanna, and Western Railroad (DL&W) was originally formed to bring anthracite coal to the Easton, Pennsylvania–Phillipsburg, New Jersey area from a field near Scranton, Pennsylvania. In 1861 the Morris and Essex Railroad, one of the early lines providing commuter service between metropolitan New Jersey and New York City, extended its tracks to Phillipsburg *via* Washington, New Jersey through New Village. Later the DL&W leased the Morris-Essex line for the primary purpose of bringing coal from Phillipsburg to the New York City area. The DL&W had a large coal terminal at Hoboken.

(2) The Morris Canal was built in 1825–1830 following the completion of the highly successful Erie Canal in 1825, which resulted in over-enthusiasm for artificial waterways. The straight line distance between Phillipsburg and Jersey City is only 60 miles whereas the canal was 106.7 miles long. The fact that Lake Hopatcong was 758 feet above the Delaware at Phillipsburg and 913 feet above sea level at the Hudson caused the canal to require 28 locks and 23 inclined planes. The latter was a device used for an abrupt change in elevation. A rise, or hill, separated the two stretches of the canal bed. A set of tracks, having a gage of 12½ feet, ran from the bed of the canal up the incline and over the brow of the hill into the next stretch. The ends of the tracks were submerged in both the upper and lower levels of the canal. A large cradle, capable of holding a canal boat, was equipped with wheels to ride upon the tracks of the plane. The boat was floated in or out of the cradle. Power to haul the cradle and its burden up the inclined plane was supplied by a turbine run by water from the higher level of the canal. The turbine activated a drum which wound up a chain attached to the cradle. One of the most impressive inclined planes was the one near Stewartsville, which had an overall rise of 100 feet for a length of 1,600 feet.

Although the Morris Canal was a major factor in the transportation of coal, pig iron, and zinc across the state for over 50 years, it was never financially successful. The peak year was 1866 when 889,220 tons of freight were moved. Starting about 1870, railroad technology had reached the point where canals like the Morris began their inevitable decline. By 1900 only 125,890 tons moved on the Morris, and the canal was abandoned in 1923. Although the canal ran near the property purchased by Edison for his cement works, it is doubtful if the proximity of this means of freight transportation was a significant factor in the choice of the New Village site. It is possible that the Edison Portland Cement Company did utilize the western portion of the canal for a time to bring in coal. (Veit,

354

R. F., "The Old Canals of New Jersey," New Jersey Geographical Press, Little Falls, New Jersey, 1963.)

(3) Dyer, F. L., Martin, T. C., "Edison, His Life and Inventions, II," pp. 511-514, Harper, New York, 1910.

(4) U.S. Patents 759, 356 (May 10, 1904), 759,357 (May 10, 1904), 802,631 (October 24, 1905).

(5) U.S. Patent 861,241 (July 23, 1907).

(6) Tricalcium aluminate has an appreciable solubility in water and hydrates very rapidly to congeal to a cement paste. Gypsum, calcium sulfate dihydrate, acts as a retarder by converting the aluminate to calcium sulfoaluminate, which is much less soluble. The optimum amount of gypsum is determined experimentally and is influenced not only by the calcium aluminate content but also by the overall fineness of the cement and its alkali content.

(7) U.S. Patent 1,219,272 (March 13, 1917).

(8) Warner, C. D., *Cement World* (June 15, 1909) **3** (3), 215. This article was also published in the *American Carpenter and Builder* (July 1909) **7** (4), 437.

(9) Halloran, Barbara, *Union Leader*, issues of February 11 and February 18, 1960.

(10) *Concrete Construction* (June, 1965) **10**, 204.

(11) Aylsworth, J. W., Dyer, F. L., U.S. Patent 1,087,098 (February 17, 1914).

(12) Sherman, Joseph, Department of Housing and Urban Development, Washington, D.C., private communication, April 1970.

(13) Kreier, Jr., G. J., U.S. Patent 3,479,786 (November 1969).

(14) *Furniture World*, "Edison Produces Concrete Furniture" (December 14, 1911).

Chapter 7 (pp. 203-233)

(1) The Lalande cell continues to be the most reliable battery for powering semaphore signals and track circuits of railroads. In 1964 a total of 1.5 million Lalande cells were manufactured in the United States. ("Kirk-Othmer Encyclopedia of Chemical Technology," 2nd ed., Vol. 3, p. 122, Interscience, New York, 1964).

(2) *Ibid.*, p. 172.

(3) Turnock, L. C., "The Active Materials and Electrolyte of the Alkaline Storage Battery," *Met. Chem. Eng.* (1916) **15**, 259-262.

(4) Holland, W. E., "The Edison Storage Battery," *Elec. World* (1910) **LV** (17), 1080.

(5) The "Kirk-Othmer Encyclopedia of Chemical Technology" (Ref. *1*) has 173 pages dealing with batteries and is an excellent reference on storage batteries. The book, "Storage Batteries," by George W. Vinal has gone through several editions. The original (John Wiley and Sons, 1924) carries a de-

tailed discussion of the Edison battery, and at this date should be representative of the early battery. Vinal's latest and 4th edition was in 1955.

(6) Salom, P. G., *Met. Chem. Eng.* (1910) 8, 633-35.

(7) Bennett, C. W., Gilbert, H. N., "Some Tests of the Edison Storage Battery," *Met. Chem. Eng.* (1913) 11, 281-82.

(8) Jones, F. A., "The Life Story of Thomas Alva Edison," p. 258, Grosset and Dunlap, New York, 1931.

(9) Josephson, Matthew, "Edison," p. 414, McGraw-Hill, New York, 1959.

(10) "Report of the Royal Ontario Nickel Commission," p. 94, A. T. Wilgress, Printer, Toronto, Ontario, Canada, 1917.

(11) Ford, Henry with Crowther, Samuel, "My Life and My Work," pp. 234-35, Doubleday, Page and Co., Garden City, 1922.

According to this reference the meeting between Edison and Ford took place in Atlantic City, New Jersey in 1897 following an address given by Edison before an "electrical meeting." In a later book by the same authors ("Edison as I Know Him," Cosmopolitan Book Corporation, New York, 1930), a different version of the first meeting between the two is given. According to this book, the meeting occurred at Manhattan Beach, Long Island, on August 11, 1896, at dinner following the technical meeting of representatives of the various Edison Illuminating Companies. Edison's comments were supposed to have been as follows: "Young man, that's the thing—you have it. Keep at it. Electric cars must keep near to power stations. The storage battery is too heavy. Steam cars won't do either, for they have to have a boiler and fire. Your car is self-contained—carries its own power plant—no fire, no boiler, no smoke, and no steam. Gasoline! You have the thing. Keep at it"!

A somewhat different version is given by Allan Nevins, "Ford: the Times, the Man, the Company," p. 167, Scribner's, New York, 1954. The meeting is described as being held at Manhattan Beach with Edison bringing his fist down on the table and exclaiming, "Young man, that's the thing! You have it—the self-contained unit carrying its own fuel with it! Keep at it!"

It would appear that the time was the August 1896 meeting of the Association of Edison Illuminating Companies. This was the first such meeting which Ford, as chief engineer of the Edison Illuminating Company of Detroit, had attended. These illuminating companies were licensed originally by the Edison Electric Light Company and had retained close ties with the parent company. However, after the General Electric merger in 1892 and the financial "panic" of 1893, the illuminating companies became largely independent. After the merger Edison continued to attend the last day of the annual meetings, primarily to meet old friends. It is likely that any "address" as mentioned by Ford was only a few remarks at the dinner which was the concluding occasion at such meetings. Edison was "knee-deep" in his iron ore project and work on the phonograph at the time and was doing little or nothing on electricity.

We must rely entirely on Ford's memory as to what the encouraging remarks were which Edison gave the young engineer at their first meeting. One would think Ford's memory of what happened in 1896–97 would have been better in 1922 than in 1930. Also, it is hard to believe how Edison could have been "carried away" by Ford's use of the internal combustion engine for power and gasoline as fuel. Edison was an ardent reader of the

References and Appendices

technical literature and was no doubt aware that such gasoline-powered cars had been built in Europe since 1885. Such a powered automobile built by the Duryea brothers had won the *Chicago Times-Herald* race held in Chicago in November 1895 in which two electric autos were among the competition. It would also appear that the comments from the 1930 book and that of Nevins would hardly be those from a man who was on the verge of starting a major research program to develop an electric battery for the automobile and also from one who at the time thought that the supply of gasoline would be inadequate for large numbers of autos and trucks so powered,

(12) Rae, J. B., "Henry Ford," Prentice-Hall, Englewood Cliffs, New Jersey, 1969. Although Ford apparently did most of his extracurricular work evenings and Saturday nights, in testimony in connection with the famous Selden patent case, he stated that he had worked on his experimental auto at Edison Illuminating, "whenever I had spare time." When asked if he had completed any autos while working for Edison, Ford replied, "Yes, two." The Fords moved to a house on Bagley Avenue in December 1893 after he became chief engineer in order to be nearer the man plant and offices of the Edison company, which were at Washington Avenue and State Street. At the rear of the lot of the double house on Bagley, there was a shed for storing wood and coal. Henry at once established an experimental workshop in his half. However, it was not until 1897 that Ford had a lathe and other substantial machinery in his private shop, but he continued to use such machinery at the Edison plant.

Chapter 8 (pp. 234-272)

(1) Miller, F. T., "Thomas A. Edison," p. 252. Printed in U.S.A., 1931.

(2) "Kirk-Othmer Encyclopedia of Chemical Technology," 2nd ed., Vol. 15, p. 177, Interscience, New York, 1968.

(3) Baekeland, L. H., *Ind. Eng. Chem.* (1909) **1**, 149.

(4) *Ibid.*, p. 545.

(5) Aylsworth, J. W., U.S. Patent 1,020,593 (March 19, 1912). Aylsworth also filed a second application in February 1910 which issued as 1,029,737. This dealt with making phenolic resins using no catalyst by slowly adding formaldehyde to phenol in an autoclave at a temperature of 260°–340°F. Only soluble resins are obtained. This patent never became commercially important. In 1911 Aylsworth also obtained British Patents 3,496-7-8, all dealing with phenolic resins.

(6) Baekeland, L. H., *Ind. Eng. Chem.* (1911) **3**, 518-520.

(7) Haynes, William, "American Chemical Industry—A History, III," p. 381, Van Nostrand, New York, 1954. Two additional patents of Aylsworth involved in this decision were 1,065,495 (1913) and 1,137,374 (1915).

(8) Redman, L. V., Weith, A. J., Brock, F. P., *Ind. Eng. Chem.* (January 1914) **6**, 3.

(9) Baekeland, L. H., *Ind. Eng. Chem.* (February 1914) **6**, 167.

357

(10) Redman, L. V., Weith, A. J., Brock, F. P., *Ind. Eng. Chem.* (1914) **6,** 263.

(11) Brown, Kirk, *Ind. Eng. Chem.* (1916) **8,** 1171.

(12) Redman, L. V., *Ind. Eng. Chem.* (1928) **20,** 1274.

(13) Aylsworth, J. W., U.S. Patents 914,222 and 914,223 (March 2, 1909).

(14) U.S. Patent 1,002,505 (September 5, 1911).

(15) Aylsworth, J. W., U.S. Patent 1,111,289 (September 22, 1913).

(16) Aylsworth, J. W., U.S. Patents 1,213,142 and 1,213,143 (January 23, 1917).

(17) Britton, J. W., Poffenberger, Noland, Midland, Michigan, private communication, September 1970.

(18) Hale, W. J., Britton, E. C., *Ind. Eng. Chem.* (1928) **20,** 114.

(19) Millikan, R. A., *Science* (January 15, 1932) **75,** 68.

(20) Josephson, Matthew, "Edison," p. 452, McGraw-Hill, New York, 1959.

(21) Bryan, G. S., "Edison: The Man and His Work," pp. 235-36, Garden City Publishing, Garden City, 1926.

(22) Edison, T. M., Llewellyn Park, West Orange, New Jersey, private communication, 1970.

(23) Compton, K. T., *Science* (January 15, 1932) **75,** 70–71.

(24) Edison, Charles, "Address before the Edison Pioneers," February 11, 1942. The occasion was a luncheon on the 95th anniversary of the birth of Thomas Edison. Charles Edison was at that time governor of New Jersey.

Chapter 9 (pp. 273-319)

(1) Firestone, Harvey, "Men and Rubber," Reinhold, New York, 1926.

(2) Davis, C. C., Blake, J. T., "Chemistry and Technology of Rubber," p. 683, Reinhold, New York, 1937.

(3) McCollum, W. B., *Ind. Eng. Chem.* (1926) **18,** 1121–24.

(4) *Ind. Eng. Chem.* (1926) **18,** 1104–1178.

(5) Polhamus, L. G., "Plants Collected and Tested by Thomas A. Edison as Possible Sources of Domestic Rubber," p. 9, U.S. Department of Agriculture, Agricultural Research Service ARS 34-74, July 1967.

(6) Prince, C. A., Jr., St. Petersburg, Florida, private communication, 1969.

(7) Ukkelberg, H. G., Richmond Hill, Georgia, private communication, December 1970.

(8) Polhamus, L. G., "Rubber Content of Various Species of Goldenrod," *J. Agr. Res.* (1933) **47** (3), 149–152.

(9) "Kirk-Othmer Encyclopedia of Chemical Technology," 2nd ed., Vol. 7, p. 676, Interscience, New York, 1965.

(10) Howard, F. A., "Buna Rubber," Van Nostrand, New York, 1947.

(11) This interdepartmental committee consisted of members of the U.S. Bureau of Standards, the Rubber Division of the Department of Commerce, and members of the Department of Agriculture. Its 57-page report dealt with the economic vulnerability of the United States if Far East rubber supplies were cut off in the event of war.

(12) Polhamus, L. G., Chevy Chase, Maryland, private communication, 1970.

(13) Staff article, "Guayule Rubber Production by the Emergency Rubber Project," *India Rubber World* (January 1944) **109** (4), 363.

(14) Eskew, R. K., "Natural Rubber from Russian Dandelion," *India Rubber World* (January 1946) **113** (4), 517.

(15) Blondeau, R., Curtis, J. T., "Latex and Rubber Characteristics of Cryptostegia," *Rubber Age* (April 1946) **59** (1), 57-63.

(16) McKennon, F. L. *et al.*, "Rubber from Goldenrod," *India Rubber World* (1948) **118, 655, 802**. This work is reported in more detail in a two-volume report of the U.S. Department of Agriculture entitled, "Extraction, Characterization, and Utilization of Goldenrod Rubber."

(17) Rollins, M. L., Bailey, Jr., T. L. W., de Gruy, I. V., "Microscopic Studies in Connection with the Extraction of Rubber from Goldenrod," *India Rubber World* (1945) **113, 75**.

(18) McKennon, F. L., Lindquist, J. R., "A Solution Method for the Compounding of Goldenrod Rubber," *Rubber Age* (1944) **56, 288**.

(19) *Rubber World* (February 1970) **161** (5), 29.

Chapter 10 (pp. 320-344)

(1) Millikan, R. A., *Science* (January 15, 1932) **75, 69**.

(2) A biography of Acheson entitled "Edward Goodrich Acheson—A Biography," has been recently written by Raymond Szymanowitz, former executive vice-president of Acheson Industries, Incorporated and a close associate of Acheson for many years (Vantage Press, New York, in press).

(3) The Hall of Fame for Great Americans, which honors U.S. citizens who have achieved lasting distinction, stands at the summit of University Heights on the uptown campus of New York University. It was established in 1900. Any man or woman who was a citizen of the United States, who made his home in the United States, and who has been deceased 25 years or more is eligible for election. The choices are entirely in the hands of its electoral college, made up of approximately 100 men and women from every state of the Union, and from every field of endeavor.

    Bronze portrait busts of those who have been elected are placed, facing one another, in the open-air colonnade. Below each bust is a recessed tablet which commemorates the person honored.

(4) Edgar, C. L., *Science* (January 15, 1932) **75, 62**.

(5) Edison, T. M., talk before M.I.T. Club of Northern New Jersey, January 24, 1969.

(6) There seems to have been a close association of employees in the various Edison ventures. For example, in 1918 Edison's employees, who had been associated with him before or during 1900, formed an organization known as the Edison Pioneers. Those employed before 1886 were full members; those from 1886–1900, associate. In 1945 the membership was expanded to include those up to and through 1899 as full members and all others up to October 18, 1931 as associates. Each year a dinner with appropriate speeches was held at the anniversary of Edison's birthday. Descendants of the Pioneers were later included in the organization. The Edison Pioneers was dissolved in 1958.

The Thomas Alva Edison Foundation was incorporated in 1946 with original headquarters at West Orange, New Jersey. It is now located at 2000 Second Avenue, Detroit, Michigan. It has a full-time executive director. The Foundation sponsors several "Edison Science Youth Days" each year and publishes and distributes literature relating to science. The Foundation has as its objective, "Advancing Science and Engineering Education."

(7) Runes, D. D., "The Diary and Sundry Observations of Thomas Alva Edison," p. 11, Greenwood Press, New York, 1968.

Index

Index

A

Acetanilide, 251
Acheson, Edward, 67, 106, 326, 336
Alan Wood Steel Co., 174
Amberol records, 129
American Association for the Advancement of Science (AAAS), 16, 46, 136
American Chemical Society (ACS), 238, 283, 321
American Graphophone Co., 108, 109
Anderson, Harold S., 101
Aniline, 250
Antisubmarine devices, 264–6
Aspdin, Joseph, 179
Aylsworth, Jonas W., 179: chlorine compounds, 256; foam concrete, 198; phenol process, 257–8; phenolic resins, 131, 238–43

B

Baekeland, Leo H., 260, 329, 336: phenolic resins, 131, 236–45
Ball, C. M., 141
Bamboo, 56–8, 64
Barker, George F., 33, 52, 104, 321
Baruch Committee, 273, 305–6, 308
Batchelor, Charles, 40, 134: electric light, 63, 66, 91; phonograph, 101, 104
Bectelsville, Pa., 147
Bell, Alexander G., 324, 334: telephone, 20, 28, 232; Volta Laboratory, 106
Belt conveyors, 161–2
Benzidine, 248
Bergmann & Co., 90, 94, 112
Bergmann, Sigmund, 89–90, 105, 227
Berliner, Emile, 110–11, 125
Bethlehem Steel Co., 169
Birkinbine, John, 142, 145, 146, 150, 168

363

D

F

G

The text of this book is set in 11-point Baskerville with two points of leading. The chapter headings are set in 11-point Baskerville Italic; the chapter titles are set in 18-point Cheltenham Bold Extra Condensed.

The book is printed offset on Sebago Antique Text, 50-pound, acid-free (pH 8.0) paper. The cover is Milbank Linen.

Book and jacket design by Joseph Jacobs.
Editing and production by Mary Westerfeld.

The book was composed by Mills-Frizell-Evans Co., Baltimore, Md., printed and bound by The Maple Press Co., York, Pa.